Ecological Studies

Analysis and Synthesis

Edited by

W.D. Billings, Durham (USA) F. Golley, Athens (USA)

O.L. Lange, Würzburg (FRG), J.S. Olson, Oak Ridge (USA)

H. Remmert, Marburg (FRG)

Volume 45

Ecological Studies

Volume 38
The Ecology of a Salt Marsh
By L. Pomeroy and R. G. Wiegert
1981. xiv, 271p. 57 figures. cloth
ISBN 0-387-90555-3

Volume 39
Resource Use by Chaparral and Matorral
A Comparison of Vegetation Function in Two Mediterranean Type Ecosystems
By P. C. Miller
1981. xviii, 455p. 118 figures. cloth
ISBN 0-387-90556-1

Volume 40
Ibexes in an African Environment
Ecology and Social System of the Walia Ibex in the Simen Mountains, Ethiopia
By B. Nievergelt
1981. ix, 189p. 40 figures. cloth
ISBN 3-540-10592-1

Volume 41
Forest Island Dynamics in Man-Dominated Landscapes
Edited by R. L. Burgess and D. M. Sharpe
1981. xviii, 310p. 61 figures. cloth
ISBN 0-387-90584-7

Volume 42
Ecology of Tropical Savannas
Edited by B. J. Huntley and B. H. Walker
1982. xi, 669p. 262 figures. cloth
ISBN 3-540-11885-3

Volume 43
Mediterranean-Type Ecosystems
The Role of Nutrients
Edited by F. J. Kruger, D. T. Mitchell, and J. U. M. Jarvis
1983. xiii, 552p. 143 figures. cloth
ISBN 3-540-12158-7

Volume 44
Disturbance and Ecosystems
Components of Response
Edited by H. A. Mooney and M. Godron
1983. xvi, 294p. 82 figures. cloth
ISBN 3-540-12454-3

Volume 45
The Effects of SO_2 on a Grassland
A Case Study in the Northern Great Plains of the United States
Edited by W. K. Lauenroth and Eric M. Preston
1984. approx. 224p. 86 figures. cloth
ISBN 0-387-90943-5

Volume 46
Eutrophication and Land Use
Lake Dillon, Colorado
By William M. Lewis, Jr., James F. Saunders, III, David W. Crumpacker, and Charles Brendecke
1984. approx. 224p. 67 figures. cloth
ISBN 0-387-90961-3

Volume 47
Pollutants in Porous Media
Edited by B. Yaron, J. Goldshmid, and G. Dagan
1984. approx. 350p. approx. 116 figures. cloth
ISBN 0-540-13179-5

The Effects of SO_2 on a Grassland

A Case Study in the Northern Great Plains of the United States

Edited by
W.K. Lauenroth and Eric M. Preston

With 86 Figures

Springer-Verlag
New York Berlin Heidelberg Tokyo

W.K. Lauenroth
Department of Range Science and
Natural Resource Ecology Laboratory
Colorado State University
Fort Collins, Colorado 80523
U.S.A.

Eric M. Preston
United States Environmental
Protection Agency
Environmental Research Laboratory
200 Southwest 35th Street
Corvallis, Oregon 97330
U.S.A.

Library of Congress Cataloging in Publication Data
Main entry under title:
The Effects of SO_2 on a grassland.
(Ecological Studies; v. 45)
Bibliography: p.
Includes index.
1. Plants, Effect of sulphur dioxide on—Great Plains.
2. Grassland ecology—Great Plains. 3. Plants, effect of air pollution on—Great Plains. I. Lauenroth, William K. II. Preston, Eric M.
QK753.S85E33 1984 574.5'2643'0978 83-20281

Typeset by Ampersand Inc., Rutland, Vermont.
Printed and bound by Halliday Lithograph, West Hanover, Massachusetts.
Printed in the United States of America.

9 8 7 6 5 4 3 2 1

ISBN 0-387-90943-5 Springer-Verlag New York Berlin Heidelberg Tokyo
ISBN 3-540-90943-5 Springer-Verlag Berlin Heidelberg New York Tokyo

Foreword

When Springer-Verlag undertook publication of this volume, two opportunities arose. The first was to bring together the significant findings of the interacting parts of a large field experiment on a whole ecosystem. Scientific specialists and the public are rightly concerned with large-scale impacts of human activity on landscapes and with the challenge of predicting subtle, long-range repercussions of air pollution. A fundamental issue is whether ecological systems like grasslands, which have evolved for several million years under stressful conditions such as variable climate and overgrazing, are more robust than other systems in tolerating new atmospheric impacts of pollution and toxicity. At what level, and when, will an extra geochemical input, like sulfur (Chapter 4), an essential nutrient for proteins and life processes, become an overload on these systems? Some grasses and grassland ecosystems seem fairly adaptable to burdens in addition to those of weather change and tissue removal. How can experts learn to project the future of the heartland of America and other grasslands of the world on the basis of only a few years of observation and control?

The second opportunity addresses a broader aspect of the project that is of interest to many readers who are not concerned with details of physiology or food chains, or the overall productivity and variations of a single plant-animal-soil community. It is the assessment of a "human experiment" which involved participants and organizations that were sensitive to the potential problems of environmental stress in the 1970s. There will be few field experiments on the scale of the one described in this volume—never enough to provide equivalent local details for every kind of grassland, much less for all other ecosystems. This and future ecological experiments will each have peculiarities of motivation and history that make them worthy of unique documentation—beyond the usual forms of hasty project reports or refined but narrow journal articles. It becomes necessary to summarize and draw on the *learning experiences* of the present case history and future sequels to integrate locally derived data and principles of specialists for broader purposes. Can we look back and find what it takes to reach beyond particular facts and conclusions?

Reports cited in Chapter 1 and elsewhere present details of early monitoring of variable conditions around an existing coal-fired electric power station at Colstrip, Montana (U.S. EPA, 1974; Lewis et al., 1975, 1976). Ecologists of the region were concerned about potential additive effects of more development in spite of the known problems of collecting statistically significant evidence under such variable conditions. It was recognized (Lee et al., 1976) that the issues of uncontrolled exposures to sulfur oxides and other contaminants in concert with varying environmental conditions could only be resolved with well-planned sampling design, analysis, and experimental control (Chapters 3, 5, and 7).

Techniques were developed for fumigation of an open grassland, but the subtleties of low-level effects in different places, seasons, and years created additional problems to be solved. Time was required to characterize the interesting differences and overall conditions of the two sites of mixed prairie which were samples of the northern Great Plains region (Chapter 2). It eventually became necessary to focus on the more vulnerable forms of life (Chapters 5 and 6). The concurrent development of the descriptive and analytical phases of the project was hampered by primary effects that turned out to be fascinating, but less dramatic than first anticipated.

One of the biggest challenges for the researchers from various organizations was to translate knowledge of the real system into abstract statements (program code) for a grassland model. An earlier volume in the Ecological Studies Series (Innis, 1977) provided a starting point for this kind of synthesis. Strategies and most of the subroutines of the model were refined (Chapter 7). For instance, cycling of several chemical elements was treated simultaneously (see Fig. 7.1, p. 163). Such geochemical refinements are likely to be followed by expanding research on models combining or linking nutrient cycles on different scales of space and time. Seasonal climatic cycles are also essential elements for many physiological and ecological predictions. Departures of day-to-day weather from climatic norms may prove unnecessary for some long-range studies, but may have to remain explicit (as in Chapters 3 to 7) in order to explain why a given pollutant release leads to readily detectable changes at some times but not at others.

The project was described as a "*multi*disciplinary" effort. As more connections between plant, animal, and abiotic parts of the landscape surfaced, scientists far from the site of the actual experiment made efforts to attain the exchange, feedback, and joint work typical of "*inter*disciplinary" discovery. From the distance of this editor, more of the latter, earlier, may have been desirable where problems transcended the individual sciences. The commitment to an interdisciplinary effort was frustrating and difficult for managers, principal investigators, and field personnel alike when the need for results in the separate disciplines and the pressure to focus on tasks and deadlines of narrower scope consumed all available time and attention. An overriding concern was, and will remain: how to ensure that the motivation and reward for such interaction can be started early, and maintained, and recognized. How can leaders guide overall planning and progress while encouraging revisions based on fresh discovery? Differences that arose between researchers were often overcome and even turned to mutual advantage. Flexibility, humor, and good will are all prerequisites in such an interdisciplinary effort.

The existence of this book, and the synthesis in Chapters 7 and 8, reflect a monumental effort to overcome the difficulties inherent in such work. The authors and editors worked long beyond the time when formal project support was over. Together they show us a mural of the prairie, with greater depth of focus and breadth of field than could be seen in just vignettes of its foreground or in a panoramic background.

Ranchers, farmers, and gardeners, as well as the scientists and technicians, who have sweated under the midday sun, can appreciate the thought, hard struggle, and patience that went into the gathering of information that finds a place in this grassland mural. To those who also struggled putting the pieces into perspective, the Editorial Board and a much wider audience of readers are further indebted. I am personally grateful to the volume editors and to Mel Dyer and Lorene Sigal for advice that helped clarify aspects of this notable experiment in orchestrating of research effort, in addition to the scientific results from that effort.

Jerry S. Olson

Oak Ridge, Tennessee
February, 1984

Preface

On April 22, 1970, a revolution was begun in the United States. Earth Day, the first national environmental teach-in, focused every eye in America on the problems of environmental degradation. The enemies were population and pollution. *The Environmental Handbook,* edited by Garrett DeBell,[1] perhaps better than any other single document, captured the mood of the moment. Air pollution received a share of the attention in this catalog of environmental problems. "There are two common denominators for air pollution from place to place: *first*, all air pollution is harmful, not only to people but to animals, plants, buildings, bridges, crops, and goods; *second*, all types are increasing everywhere at such a rate that researchers are left gasping" (DeBell, 1970, page 120, excerpt by Robert and Leona Train Rienow from their book *Moment in the Sun,* 1967, Sierra Club—Ballantine Books).

This wave of environmental concern was not limited to the United States. In March, 1970, the International Social Science Council Standing Committee on Environmental Disruption convened an international symposium in Tokyo, Japan. In May, 1969, U Thant, then United Nations Secretary General, proposed to hold an International Conference on Human Environment in June, 1972, in Stockholm, Sweden. The Stockholm Conference was a benchmark in arousing worldwide concern for the environment.

---

[1] DeBell, G. 1970. *The Environmental Handbook.* Ballantine Books, New York.

Understanding this revolution simultaneously provides an explanation for why this project was begun in the early 1970s and an indication of the pressures influencing public and scientific thinking about air pollution at this time. Planning for a project on the bioenvironmental impacts of coal-fired power plants was begun in 1972–73, and the first experiments were begun in 1974. The project was initiated by the United States Environmental Protection Agency (EPA) through the National Ecological Research Laboratory (now the Environmental Research Laboratory) in Corvallis, Oregon. Project participants were drawn from the EPA, Colorado State University, Montana State University, and the University of Montana. The United States Department of Agriculture (USDA) was involved in the project as the managers of land used for the experiments (Custer National Forest). More than a dozen senior scientists and their graduate students and technicians were involved in the project. The objective of this book is to summarize and interpret the results of this effort.

We chose a format of presentation to emphasize the generalities of our findings. The details of the experiments can be found in the published reports to which we refer. In Chapter 1 we present an introduction and overview of the problem and the rationale behind our approach. Chapters 2 and 3 are descriptions of the study sites and the SO_2 exposure system. Chapters 4–6 summarize and interpret the major findings from the field experiments. These findings are incorporated into simulation modeling exercises in Chapter 7. As the experiments progressed, hypotheses were formed and tested, then supported, rejected, or revised. New hypotheses were formed and the cycle repeated. Thus, there is not a continuous data set for most measurements over the entire course of the experiment. We attempted to weave key pieces of information into an overall view of the effects of SO_2 exposure on the experimental site. Chapter 8 interprets the results in the context of potential impacts of energy development in the northern Great Plains, other grasslands, and other vegetation types.

This book represents the efforts and hard work of many persons. To all of them we are grateful. For their intellectual and leadership contributions, we thank J.J. Bromenshenk, J.D. Chilgren, J.K. Detling, J.L. Dodd, J.E. Taylor, N.R. Glass, S. Eversman, C.C. Gordon, W. Grodzinski, J.E. Heasley, J.J. Lee, J.W. Leetham, A.S. Lefohn, R.A. Lewis, D.G. Milchunas, M.L. Morton, W.J. Parton, D.M. Swift, D.T. Tingey, P.C. Tourangeau, and R.G. Woodmansee. We thank those who labored so faithfully in the field and laboratories: T.L. Gullett, T. Fletcher, R.K. Heitschmidt, T.J. McNary, P.M. Rice, L. Renerio, W.C. Leininger, C. Liggon, R. Fuchs, D.E. Body, G.L. Thor, D.B. Weber, C.J. Bicak, W.S. Ferguson, S.R. Bennett, M.B. Coughenour, and R.B. Deblinger.

The experimental sites were located on public land administered by the USDA Forest Service. We benefited greatly from the cooperation of Forest Service personnel, especially R. Meinrod.

The publications sections at the Environmental Research Laboratory (EPA) and the Natural Resource Ecology Laboratory (Colorado State University), especially J. Hill, S. Taylor, E. Taylor, and K. Curry, deserve special thanks for their efforts.

The final version of the manuscript benefited greatly from the review comments of J.S. Olson, P.G. Risser, D.T. Tingey, F.E. Clark, and D.C. Coleman.

Financial support for the field experiments was provided by the Environmental Protection Agency and the Colorado State Agricultural Experiment Station. A substantial portion of the simulation modeling work was supported by the Electric Power Research Institute.

Contents

Contributors

J. L. Dodd — Range Management Division
University of Wyoming
Laramie, Wyoming, U.S.A.

J. E. Heasley — Private Consultant
Formerly Natural Resource Ecology Laboratory
Colorado State University
Fort Collins, Colorado, U.S.A.

T. B. Kirchner — Natural Resource Ecology Laboratory
Colorado State University
Fort Collins, Colorado, U.S.A.

W. K. Lauenroth — Range Science Department and
Natural Resource Ecology Laboratory
Colorado State University
Fort Collins, Colorado, U.S.A.

Jeffrey J. Lee — United States Environmental
Protection Agency
Corvallis, Oregon, U.S.A.

J. L. LEETHAM
Private Consultant
Formerly Natural Resource Ecology Laboratory
Colorado State University
Fort Collins, Colorado, U.S.A.

D. G. MILCHUNAS
Range Science Department and Natural Resource Ecology Laboratory
Colorado State University
Fort Collins, Colorado, U.S.A.

ERIC M. PRESTON
United States Environmental Protection Agency
Corvallis, Oregon, U.S.A.

T. P. YORKS
Natural Resource Ecology Laboratory
Colorado State University
Fort Collins, Colorado, U.S.A.

1. Potential Effects of SO_2 on the Northern Mixed Prairie: Overview of the Problem

ERIC M. PRESTON AND W. K. LAUENROTH

1.1 Introduction

Semiarid grasslands in general and those of the Great Plains of North America in particular have been subjected to a wide variety of uses ranging from moderate domestic herbivore grazing to strip mining or plowing for crop production. The demands for food and fuel in the next several decades will place increasing pressure on the Great Plains grasslands. The objective of this study was to investigate the role that sulfur dioxide air pollution might play in causing perturbations to grasslands of the northern Great Plains as the substantial coal reserves of the region are developed.

1.2 Characteristics of Grasslands

The structure and behavior of all ecological systems vary continuously in both space and time. The surrounding environment is ever-changing. Some of the grassland areas in the northern Great Plains were under a sheet of ice less than 20,000 years ago and may have cycled between forest and grassland types in the time between. Climate also influences topography and soils through direct action as well as through vegetational and animal activity. These changes in turn feed back to system structure and behavior. Too often, humans are portrayed as the only vector of change in ecological systems. Often, when we act it is to influence the magnitude and/or direction of changes already in progress.

For shorter time spans, season to season and year to year, changes in characteristics of ecological systems are the rule rather than the exception. These changes are not random shifts in structure or behavior; rather, they represent adjustments to an environment that is only partially predictable. These variations are empirical evidence of the multitude of feedback loops between an ecological system and its environment. According to this interpretation, the multiple dimensions of system structure and function are continuously regulated by the multiple dimensions of the abiotic and biotic environment, including human manipulations and perturbations resulting from our activities.

Holling (1978) interpreted this dynamic variability as a key to understanding how ecological systems respond to and persist in the face of perturbations. He defined a term, resilience, to describe in a qualitative sense the characteristic of ecological systems which enables them to adjust to changes and yet persist in fundamentally the same form. Resilience then is an indication of the number of alternative states in which a system can exist and still be recognizable as a forest or a grassland or whatever its state was before the disturbance. The amount of variability in the environment during the past development of the system is suggested by Holling to be an important determinant of its ability to adjust to current perturbations. Thus, ecological systems that developed under fluctuating environments are expected to have a greater capacity to adjust to perturbations than those that developed under constant conditions.

Holling (1978) cites information on forest insects throughout Canada from Watt (1968) to support this assertion. Populations of insects from maritime regions were reported to be more sensitive to fluctuations in air temperature than were those from continental regions. Grasslands in North America also lend support to this model. Throughout the central grassland region environmental variability is high (Weins, 1974). In many instances the variance of precipitation or air temperature provides more information than the mean. The ability of these grasslands to adjust to and recover from heavy grazing by domestic livestock and drought are well documented (Weaver and Albertson, 1944; Albertson et al., 1957). Additional evidence of the ability of these grasslands to adjust to perturbations can be found in manipulation experiments (Frayley, 1971; Lauenroth et al., 1978).

In addition to specific environmental results documenting resilience of grasslands, we find broad evidence for the hypothesis that natural grasslands in general possess a high degree of resilience. Humankind's immense success in establishing the food-energy base of civilization in the grassland regions of the world by trial-and-error management is evidence of the degree of disturbance these systems can withstand.

Does this mean that air pollution should have a greater or lesser effect on grasslands than other types of ecological systems? If grasslands have a capacity to withstand air pollution impacts, what characteristics impart that capacity? The following chapters address these and other questions about the impacts of sulfur dioxide air pollution on grasslands.

The scope of our approach to the impact of SO_2 exposure on a northern mixed prairie ranged from the microscopic to the macroscopic. At the lowest levels of organization (greatest detail), we examined the cellular structure of lichens. At the highest level of organization, we evaluated net production of the vegetation and the structure of the plant and animal (invertebrate) communities. Methods used to evaluate the significance of our findings ranged from statistical analyses of the responses of individual system components to experiments with a simulation model.

The overall objective remained intact throughout the duration of the project. The research plan and operational objectives, on the other hand, were continually adjusted and tuned as the project progressed and we gained new information and either reformulated hypotheses or developed new ones. This resulted in the actual work accomplished being slightly different from that initially proposed.

1.3 Challenge of Coal Conversion

Approximately 40% of the minable coal reserves in the United States are located beneath the grasslands and croplands of the northern Great Plains (Northern Great Plains Resources Program, 1975) (Fig. 1.1). The low sulfur content and relatively easy access to these reserves by strip mining assures that

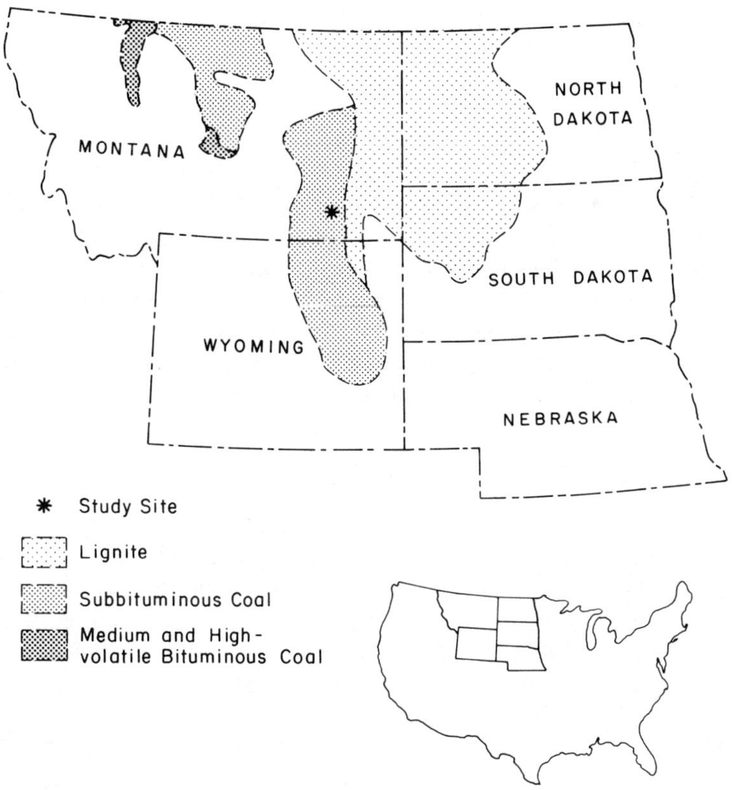

Figure 1.1. Map of U.S. coal fields underlying the northern Great Plains.

this region will be a focal point for some time of investigations of the effects of both mining and combustion of coal on the environment.

Coal development is likely to conflict with more traditional land uses. Some land will be removed from agricultural production directly by mining and combustion facilities, and the productivity of much of the remaining land could be affected by toxic emissions from these facilities. Data from permit applications (Durran et al., 1975) projected an increase in electric generating capacity from approximately 3000 Mw in 1976 to 17,000 Mw in 1986. Such a large increase in coal combustion in an area with previously near-pristine air quality raised many questions concerning the effects of air pollutants on the biota of the region. In several ways, the appropriate questions are novel and have not been previously addressed.

Native perennial grasslands are the dominant vegetation in the northern Great Plains, and livestock grazing has been the predominant agricultural use. The economic survival of the livestock industry in this region is dependent upon the long-term persistence and productivity of these grasslands. Even more than for annually cultivated systems, subtle effects of toxic emissions that are perhaps insignificant in the short term could cause cumulative problems in the long term. Under certain conditions, sulfur dioxide, for example, can act as both a toxic agent and a nutrient. The transition from nutrient to toxicant is related to the concentration, duration, and frequency of exposure.

1.3.1 Coal Reserves

The region is rich in energy resources including coal, oil shale, uranium, oil, gas, and geothermal energy (Northern Great Plains Resources Program, 1975). Of these, coal is by far the most abundant, with estimates of 1.5 trillion tons within the northern Great Plains region. Approximately 160 billion tons of this total amount are commercially minable by virtue of their seam thickness and accessibility. This represents 37% (by weight) of the known minable coal reserves in the United States. Half of this can be surface stripped; the other half will require underground mining. Refinement technology reduces the maximum net coal yield to 100 billion tons. This could produce 2000 quadrillion BTU; about 30 years of the energy consumption of the United States (U.S. Congress/Office of Technology Assessment, 1979).

Coal deposits are dispersed over the entire region, with major coal fields in eastern Montana and Wyoming and the western portion of North Dakota. Measured coal bed thickness has been reported to be as great as 60 m. The quality of northern Great Plains coal differs in several aspects from coal in the eastern United States; on the average sulfur content and heat value are lower and the water content is higher.

The patterns and rates of development of the northern Great Plains coal depend on a wide range of factors such as energy demand, price, availability of capital, availability of equipment and manpower, environmental policies and regulations,

Table 1.1. Projections for Coal Mining in Northern Great Plains (10^6 Tons)

Estimate Level	1985	2000	Reference
Low	290	490	U.S. Congress/Office of
High	320	700	Technology Assessment (1979)
Low	108	116	Montana University Coal
Intermediate	137	214	Demand Study Team (1976)
High	173	260	
Low	108	144	Northern Great Plains
Intermediate	192	362	Resources Program (1975)
High	382	977	

competing land use, and transportation. Because there are so many important variables that can only be speculatively quantified, forecasting coal development is difficult. However, because of the recent interest of both the federal government and the energy industry in coal development in this region, production is expected to increase rapidly.

Several projections of coal development are presented in Table 1.1. The major reason for the disparity among projections is the difference in assumptions used in the analyses, particularly with respect to demand for northern Great Plains coal for synthetic fuel production. In any event, speculation indicates that coal production will increase to three to five times the 1977 production by the year 2000, and coal consumption is expected to increase by two to three times (Power, 1975).

1.4 Adaptive Research

Between 1974 and 1980, the U.S. Environmental Protection Agency (EPA) supported a study to gain a portion of the information needed to address the environmental problems facing the northern Great Plains. This volume is a product of that study and it reports on a multidisciplinary field experiment designed to evaluate the consequences of sulfur dioxide exposure to a typical native grassland in the northern Great Plains. As conceived in 1974 (U.S. EPA, 1974),

> The broad objective of this program is to measure and predict change in a grassland ecosystem as a function of meaningful environmental parameters including air pollutants. We are concerned not only with the stability of ecosystem organization in relation to ambient conditions, but also with the predictability and reproducibility of changes that do occur. Insight into the mechanisms of dynamic-structural responses to air pollution challenge is also sought. It is particularly important to identify the subsystem functions that contribute to ecosystem regulation and the mechanisms whereby such regulation is effected.
>
> Based on this comprehensive investigation, we hope to generate a defensible, sensitive, and relatively simple program that may be used in other grassland situations to monitor, evaluate, and predict bioenvironmental effects of air pollution from fossil fuel conversion

processes. We envision an evolving program that will allow managers to gradually refine cost-benefit determinations in making decisions concerning site selection and air pollution control of coal conversion facilities.

1.4.1 Strategy

Staff papers written during the program's conception and planning stages suggested that a simulation model (ELM) developed during the Grassland Biome project of the U.S. International Biological Program (Innis, 1978) would be updated to provide an integrative focus for data collection and for interpretation of the effects of air pollution on energy flow and nutrient cycling. "By studying a rather broad range of interacting variables and, in particular, by an intensive study of certain populations, some may be isolated as sensitive and reliable measures of air pollution" (U.S. EPA, 1974).

Cumulative indirect effects were expected to be important in the response of ecological systems to SO_2. Therefore, in addition to studying those components likely to be especially sensitive to direct effects, system processes that could demonstrate secondary and subtle cumulative effects were studied.

1.4.2 Tactics: The Field Experiment

The primary field experiment involved exposing 0.5 hectare (ha) plots of native grassland to measured SO_2 concentrations for several field seasons. A field exposure system was developed to simulate the temporal and spatial patterns of SO_2 exposures that a grassland might experience if a significant number of coal combustion and conversion facilities were developed in the northern Great Plains.

The general objective for the project was first to document and ultimately to predict change in a northern mixed prairie as a function of environmental variables, including air pollution. It was envisioned to be a total system project. However, we expected to study representative items from each major system component rather than exhaustively study every detail of the entire system. Simulation modeling at a variety of levels of biological organization was to provide integration and synthesis of individual experimental results.

The initial plan called for a 3-year field effort with a fourth year devoted to analysis and evaluation of results. The major benefit from the project was expected to take the form of a body of information that would aid planning of energy development in the northern Great Plains.

1.4.3 Evolution of Objectives and Research Plan

The original plan called for a 4-year project. For a variety of reasons, it continued for 7 years. The planned 3-year field experiment continued for 5 years.

A full year of development of the field exposure system was required before the experiment could begin. A year of analysis followed the end of the field experiment.

Simulation modeling was initially envisioned as an integral part of the total approach and as a tool to be helpful in evaluating results and planning subsequent experiments. This initial plan was not realized. The modeling activities were used largely for analysis rather than for planning.

At the start of the experiment, objectives and hypotheses about responses of the grassland to SO_2 exposure reflected our expectation that the impacts would be quite substantial. The exposure concentrations were chosen to provide a gradient of effects from barely detectable to large and obvious. We expected to see changes in species composition of plant and animal populations. We hypothesized that primary production would be sharply reduced by exposure to our highest SO_2 concentration. We planned to evaluate a number of populations as potential bioindicators of SO_2 exposure.

As the experiment progressed, it became clear that the large and obvious impacts of SO_2 exposure, predicted from our initial hypotheses, were not being realized. Our problem was not how to choose the best among a variety of bioindicators, but whether we were even evaluating appropriate indicators of impact. In striving to answer this question, our objectives and hypotheses covered a large variety of system components often at several levels of organization.

1.4.4 Experimental Design and Analysis

The initial experimental design called for three replications of three SO_2 treatments and a control (Lewis et al., 1975; 1976). That design was not achieved but rather two replications of the treatments were utilized. The first replication was initiated in 1975 (Site I) and the second in 1976 (Site II). The details of each site are presented in Chapter 2 and a description of the SO_2 delivery system is provided in Chapter 3.

Early in the experiment we recognized several statistical problems with the design. Two replicates do not result in very powerful analyses for treatment effects. More than two replications of the experiment were operationally impossible. Additionally, many of the experiments could not be accomplished on both sites. Replication of subplots within the large experimental plots was employed for these experiments. These were located randomly and were expected to increase the precision in detecting responses as a result of SO_2 exposure. Details of the specific analyses can be found in the chapters on results or in the publications cited.

After testing a prototype exposure system and considerable discussion among project participants, the objectives for the exposure concentrations were set at 52, 130, and 260 $\mu g\ SO_2 \cdot m^{-3}$ geometric means over the growing season for the three SO_2 treatments. Daily 3-hr peaks were predicted to be 1000 to 1300 $\mu g \cdot m^{-3}$ for the highest concentration and 200 to 260 for the lowest concentration. We anticipated peaks during the growing season as high as 10,000 $\mu g \cdot m^{-3}$ for the highest SO_2 treatment (Lee et al., 1976).

Chapters 4, 5, and 6 contain the results and interpretations from the field experiment. These are presented in ecological rather than chronological order. Chapters 7 and 8 are syntheses of the results. Chapter 7 reports and interprets experiments conducted with a simulation model. In Chapter 8 we attempt to place our results into perspective for the northern Great Plains of North America, other grasslands, and other types of ecological systems.

References

Albertson, F. W., G. W. Tomanek, and A. Riegel. 1957. Ecology of drought cycles and grazing intensity on grasslands of central Great Plains. *Ecol. Monogr.* 27:27–44.

Durran, D. R., M. J. Meldgin, Mei-Kao Liu, T. Thoem, and D. Henderson. 1979. A study of long range air pollution problems related to coal development in the northern Great Plains. *Atmos. Env.* 13:1021–1037.

Fraley, L., Jr. 1971. *Response of shortgrass plains vegetation to chronic and seasonally administered gamma radiation.* Unpublished doctoral thesis, Colorado State University.

Holling, C. S., ed. 1978. *Adaptive Environmental Assessment and Management.* International Series on Applied Systems Analysis No. 3. New York: Wiley.

Innis, G. S., ed. 1978. *Grassland Simulation Model.* Ecological Studies 26. New York: Springer-Verlag.

Lauenroth, W. K., J. L. Dodd, and P. L. Sims. 1978. The effects of water and nitrogen induced stresses on plant community structure in a semiarid grassland. *Oecologia* (Berl.) 36:211–222.

Lee, J. J., R. A. Lewis, and D. E. Body. 1976. The field experimental component: Evolution of the Zonal Air Pollution System. In *Bioenvironmental Impact of a Coal-Fired Power Plant*, R. A. Lewis, A. S. Lefohn, and N. R. Glass, eds., pp. 188–202. Second Interim Report. Corvallis, OR: EPA 600/3–76–013.

Lewis, R. A., A. S. Lefohn, and N. R. Glass. 1975. An investigation of the bioenvironmental effects of a coal-fired power plant. In *Fort Union Coal Field Symposium*, W. F. Clark, ed., pp. 531–536. Billings, MT: Montana Academy of Sciences.

Lewis, R. A., A. S. Lefohn, and N. R. Glass. 1976. Introduction to the Colstrip, Montana, coal-fired power plant project. In *Bioenvironmental Impact of a Coal-Fired Power Plant*, R. A. Lewis, A. S. Lefohn, and N. R. Glass, eds., pp. 1–13. Second Interim Report. Corvallis, OR: EPA 600/3–76–013.

Montana University Coal Demand Study Team. 1976. *Projections of Northern Great Plains Coal Mining and Energy Conversion Developments 1975–2000 A.D.*, Summary Volume. Washington, DC: Research Report, NSF/Rann Grant No. AER 75-25278.

Northern Great Plains Resources Program. 1975. *Effects of Coal Development in the Northern Great Plains*. Billings, MT: Staff Report.

Power, T. M. 1975. *Electric Energy Demand and the Demand for Northern Great Plains Coal.* Montana University Coal Demand Study Working Paper No 1. Missoula, MT: Department of Economics, University of Montana.

U. S. Congress/Office of Technology Assessment. 1979. *The Direct Use of Coal: Prospects and Problems of Production and Combustion*. OTA–E86. Washington, DC.

U. S. Environmental Protection Agency. 1974. *The Bioenvironmental Impact of Air Pollution from Fossil-Fueled Power Plants*. Corvallis, OR: EPA 600/3–74–011.

Watt, K. E. F. 1968. A computer approach to analysis of data on weather, population fluctuation, and disease. In *Biometeorology*, W. P. Lowry, ed., pp. 145–159. Corvallis, OR: Oregon State University Press.

Weaver, J. E., and F. W. Albertson. 1944. Nature and degree of recovery of grasslands from the great drought of 1933 to 1940. *Ecol. Monogr.* 14:393–479.
Wiens, J. A. 1974. Climatic instability and the ecological saturation of bird communities in North American grasslands. *The Condor* 76:385–400.

2. The Plains Region and Experimental Sites

W. K. Lauenroth, J. L. Leetham, D. G. Milchunas, and J. L. Dodd

2.1 The Northern Great Plains

The northern Great Plains extend from the valley of the Platte River in Nebraska to well beyond the Canadian border and from the Rocky Mountains on the west to the Central Lowlands on the east (Fenneman, 1931). The eastern and southern boundaries are difficult to define because changes in relief, climate, and vegetation are gradual. Much of the region is underlain by coal deposits (Figure 1.1).

Land in the area can be classified broadly into four cover types. The native northern mixed prairie (rangeland and pasture) is by far the most extensive (Singh et al., 1983). Cropland, pine woodlands, and riparian communities account for successively smaller proportions of total land cover (Table 2.1).

Beef and wheat dominate agricultural production in the region. In Wyoming, Montana, and the western half of North and South Dakota, 70% of the total land area is used for pasture and rangeland, and 26% of the land is cultivated (Northern Great Plains Resource Program, 1975). Wheat is the single-most important crop, accounting for 30% of total cropland. Other dryland crops grown in the area are barley, flax, rye, and oats. Irrigated land growing mostly corn, alfalfa, and sugar beets accounts for only 3% of the total cropland. The southern half of the northern Great Plains is dominated by grazing lands and nearly 80% of the farm income is derived from livestock. By contrast, in the northern half of the region, more than 60% of total farm income is derived from wheat and other grains.

Table 2.1. Land Use Summary for the Northern Great Plains by State[1]

	Pasture and Rangeland		Cropland		Forest and Woodland		Urban	
State	Million ha	% of Total	Million ha	% of Total	Million ha	% of Total	Million ha	% of Total
Montana	10.7	76.3	2.5	18.1	0.7	4.8	0.1	0.8
North Dakota	4.2	4.4	6.0	56.9	0.1	0.6	0.2	2.1
South Dakota	3.9	82.6	0.7	15.7	0.1	1.3	0.0	0.3
Wyoming	6.5	85.6	0.3	3.7	0.7	9.1	0.1	1.6
Total	25.3	68.8	9.5	25.9	1.6	4.0	0.4	1.3

[1]Adapted from Northern Great Plains Resource Program (1975).

2.1.1 Climate

The climate of the northern Great Plains is temperate and semiarid. Mean annual precipitation ranges between 300 and 500 mm with rainfall occurring predominantly in April, May, and June. The Rocky Mountains influence the four weather-producing factors (temperature, air pressure, wind, and moisture) resulting in long, cold winters, short but warm summers, a large diurnal range of temperature, frequent strong winds, and limited and highly variable precipitation. In some years the weather is dry and even arid, and in other years it is relatively wet. Sudden temperature changes, extreme cold, severe gales, snow and blizzards in the winter, and thunderstorms, and tornados at other seasons of the year, are common.

The fluctuations of the weather create serious risks to agriculture and, to a large extent, determine the nature of native biota. The uncertainty of precipitation, the ever-present danger of drought, extremes of temperature, unseasonal frosts, and high wind velocity are among the major crop hazards of the northern Great Plains.

The climatic elements of the Great Plains vary along north–south and east–west gradients. These patterns give rise to the different types of grasslands occurring there. From east to west, precipitation and relative humidity decrease, while solar radiation, rainfall variability, water stress, and potential evapotranspiration increase. In a south to north direction, there is a decrease in air temperature, number of frost-free days, potential evapotranspiration, fraction of the precipitation occurring in the summer, solar radiation during the winter, and water stress (in the west). Both temperature and precipitation are most variable in the western and southern portions of the northern Great Plains.

2.1.2 Vegetation

The northern mixed prairie (Figure 2.1) is dominated by a mixture of short and mid-height perennial grasses (Singh et al., 1983). Cool-season perennial grasses are dominant. Warm-season shortgrasses are important secondary species

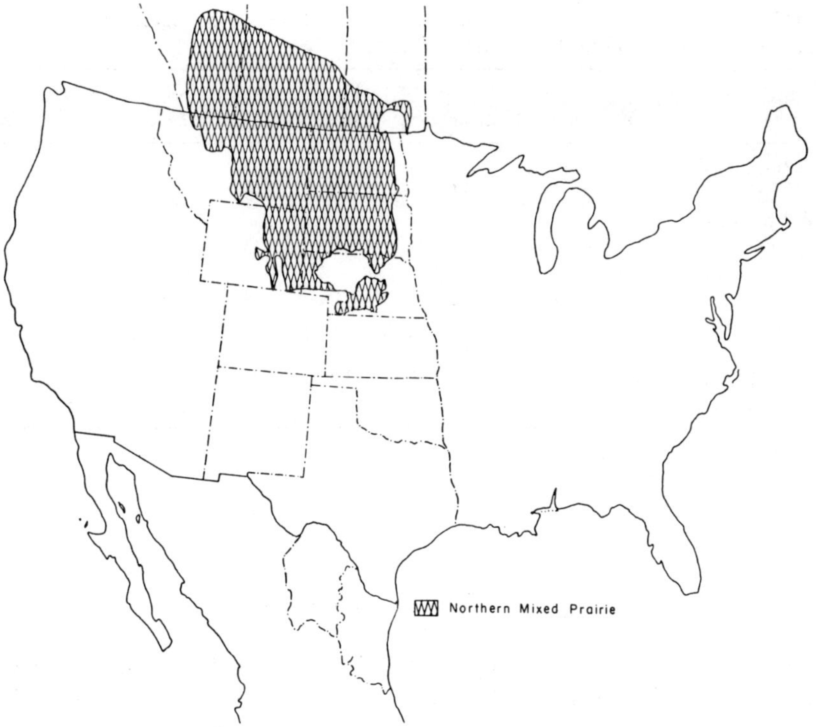

Figure 2.1. Map of northern mixed prairie.

throughout and are often codominant with the mid-height perennials. The exact blend of short and mid-height grasses at a given locale is a complex function of topographic position, soils, recent weather, and grazing history. Drier than normal springs, heavy grazing, upland locations, and fine textured soils are conditions that generally favor the shortgrasses, while the opposite conditions shift the blend more in favor of the mid-height grasses. Forbs (herbs), half-shrubs (suffrutescent), succulents, and mid- and tall warm-season grasses are persistent but rarely dominant components of the northern mixed prairie. Lichens, mosses, club mosses, and algae often occur at the soil surface.

Variation from location to location in the composition of the northern mixed prairie is considerable and usually associated with variation in abiotic characteristics such as soils, slope, aspect, and, on a broader scale, climatic variation. An example of this variation in structure is given in Table 2.2 for several reference sites. The Matador Site is in Canada, the Hay Coulee, Kluver West, Kluver North, and Kluver East sites are in southeastern Montana, and the Dickinson and Cottonwood Sites are in North and South Dakota, respectively. Within this small set of examples the contribution of cool-season grasses ranges from 47 to 94% of aboveground standing crop and the composition of half-shrubs ranges from a trace to 30%. Variation in structure similar to that demonstrated in Table 2.2 for the northern mixed prairie occurs in all grassland types.

Table 2.2. Functional Group Composition (Percentage of Aboveground Standing Crop) of Selected Sites within Northern Mixed Prairie[1]

	Dickinson	Cottonwood	Hay Coulee	Kluver West	Kluver North	Kluver East	Matador
Cool-season grasses	60	81	63	86	47	58	94
Warm-season grasses	14	15	15	3	12	5	—
Cool-season forbs	4	2	10	8	11	9	2
Warm-season forbs	4	+	+	3	+	+	—
Half-shrubs	18	+	12	+	30	28	4
Succulents	+	1	+	+	+	+	—

[1]From Singh et al. (1983).

2.1.3 Heterotrophs

Saprophages constitute a large portion of the total heterotrophic biomass and process 90–97% of the total net primary production in most terrestrial systems (Coleman et al., 1976). For the mixed prairie in Canada, the maximum standing biomass of soil microorganisms is approximately half that of the maximum standing crop of vegetation (Paul et al., 1979). Bacterial biomass is about one quarter to one half that of fungi.

Grasshoppers and other species of Orthoptera are the most conspicuous invertebrates of northern Great Plains grasslands and exert their greatest effect as aboveground herbivores. Less conspicuous belowground herbivores such as plant parasitic nematodes and soil arthropods process a greater quantity of energy than aboveground herbivores. On sites grazed by cattle, aboveground herbivory is a larger proportion of aboveground primary production than belowground herbivory is of belowground primary production.

Vertebrate species diversity in grasslands is greater than generally appreciated since many species are nocturnal, cursorial (running), or fossorial (burrowing). Common mammals include the deer mouse, prairie vole, black-tailed prairie dog, white-tailed jack rabbit, striped skunk, coyote, mule deer, pronghorn antelope, and domestic cattle. Representative birds include western meadowlark, horned lark, sage grouse, kestrel, red-tailed hawk, great horned owl, lark bunting, vesper sparrow, and savannah sparrow.

Large herbivores may be as important as climate and soil in determining floristic composition (Coupland, 1979; Mack and Thompson, 1982) since herbaceous species are not equally sensitive to grazing pressure. Browsers may restrict the growth of trees and shrubs.

2.2 Site Selection

The experimental design (Chapter 1) required that we locate a pair of sites each approximately 10 ha. The criteria for choosing the sites were that they be similar, representative of the grassland vegetation of the region, and reasonably close to each other. The remainder of this chapter describes the two sites, discusses their similarities and differences, and evaluates how representative they are of the grasslands in the region.

2.2.1 Location, Geology, and Soils

The paired experimental sites are located in southeastern Montana (45° 29′ N, 106° W) in the Fort Howes District of Custer National Forest. The sites are approximately 210 km southeast of Billings, Montana, and 110 km northeast of Sheridan, Wyoming. The topography is generally rolling to hilly with deeply dissected ridges and valleys separating uplands and lowlands. Locally, the major drainage is north via Otter Creek, with minor drainage flow, east and west. The sites are on uplands at 1200 m elevation, on southwest-facing slopes of less than 4° slope.

Throughout Montana, North Dakota, and Wyoming the predominant geologic substrate is the Fort Union formation consisting primarily of weakly unconsolidated sedimentary rocks deposited by freshwater streams during the Paleocene. The study areas lie directly over this formation, which is composed of nearly level beds of sandstone, silty sandstone, clay, and silt shale with lenses of lignite. Parent materials for soils at both sites were derived through the erosion of nearby buttes and ridges. Both the experimental soils are typic argiboralls, which, along with aridic argiboralls, are the major soil groups throughout the northern Great Plains (Aandahl, 1972).

The dominant soil type of Site I is a Farland silty clay loam. Farland soils are generally deep, well drained, and medium in texture. They are often associated with Cabba soils, which occur on steep slopes and are shallow and medium in texture. The Farland–Cabba association occurs on approximately 1% of the local region. An average profile of Site I has a solum depth greater than 100 cm and a B horizon 66 cm thick (Table 2.3). Textures range from silt loam in the A horizon to clay loams in the B horizon. The Farland series is widely distributed throughout Montana, North Dakota, and Wyoming. It is closely related to many of the other important soil series in the region.

Soils at Site II are Thurlow clay loams, which are deep, light colored, and moderately fine in texture. The most frequently associated soil is the Midway series, which is moderately fine in texture, calcareous, and generally shallow over clayey, silty, and sandy shale. The Midway–Thurlow association occurs on approximately 4% of the local region. The Thurlow soil at Site II has a 100-cm-deep solum, a 12-cm-deep A horizon, and an 88-cm-deep B horizon (Table 2.3).

Table 2.3. Selected Physical and Chemical Characteristics of the Soils at the Experimental Sites

Site	Horizon	Depth (cm)	Sand (%)	Silt (%)	Clay (%)	Organic Matter	pH	Lime (%)	N (%)	Total P (%)
I	A	0–25	61	22	17	1.0	6.1	0.0	0.064	0.029
	B	26–91	40	30	27	0.6	6.9	0.2	0.052	0.039
	C	92–134+	40	24	36	0.4	8.0	2.9	0.030	0.036
II	A	0–12	32	35	32	3.0	6.4	<0.1	0.181	0.058
	B	13–100	39	29	32	1.2	7.7	1.0	0.091	0.050
	C	101–152+	38	32	30	1.4	8.0	4.8	0.080	0.060

The A horizons were loam and clay loam in texture; the B horizon was clay loam throughout. The Thurlow series is of minor importance in the northern Great Plains.

Soil chemical characteristics reflect the differences in soil texture between the two sites (Table 2.3). The higher silt and clay contents of the soils at Site II are associated with greater organic matter and more nitrogen and phosphorous. The higher lime contents of Site II soils are consistent with higher pH measurements.

Soil water dynamics in northern Great Plains grasslands is very closely tied to precipitation patterns. Growing season evapotranspiration is sufficiently large so that no carry-over of soil water occurs from one growing season to another. On the average throughout the mixed-prairie region, soil water content is highest in April and May and lowest in August and September. The seasonal trends we observed conformed to this regional pattern; soil water was high from the beginning of the growing season until mid- to late June, declined rapidly during July, and leveled off and remained essentially constant during August and September (Figure 2.2). While this general pattern was followed, differences between years were observed. In 1975 and 1977, soil water contents were near field capacity early in the growing season and near the wilting point in August and September. In 1976 and particularly 1978, soil water contents early in the season were relatively low. Field capacity in 1976 was never realized. Approximately 300 mm of precipitation were received between early May and late June 1978. Soil water contents should have been near field capacity during this period, although confirming measurements were not performed.

2.2.2 Weather

The general characteristics of the climate of the region are conveniently illustrated using a Walter-type climatic diagram (Figure 2.3) (Walter, 1979). Winters are cold and dry. Mean monthly temperatures from November through March are near or below 0°C. Winter precipitation, largely in the form of snow, is less than 15 mm per month. Temperatures rise rapidly in April and reach maxima of greater than 20°C in July and August. Maximum precipitation usually occurs in May and June. Mean monthly temperatures fall rapidly after August.

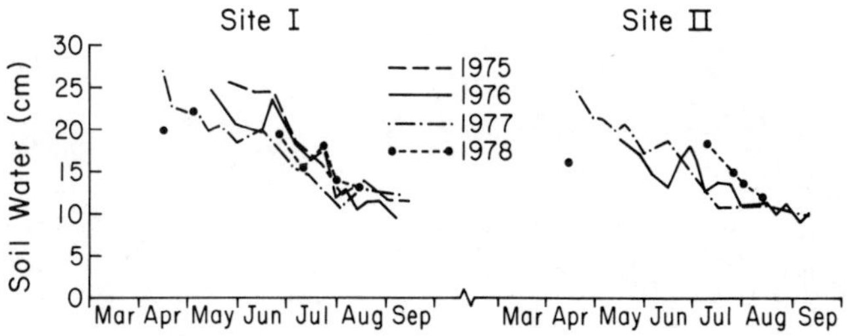

Figure 2.2. Soil water (cm) dynamics for Sites I and II 1975–1978.

Annual precipitation for the study sites is approximately 400 mm per year. Because of the complexity of the local topography, it is difficult to estimate exact site precipitation from data obtained from nearby stations. At Sonnette, 15 km north of the study sites, the average May-through-September precipitation is 235 mm, with a standard deviation of 48 mm. Maximum and minimum precipitation for May–September during the 16 years of available data were 320 and 137 mm, respectively. Data collected for this period over our 5 years of study ranged from a low of 170 mm in 1979 to a high of 415 mm in 1978. The low value is approximately 1 standard deviation below the mean from Sonnette, and the high value is almost 4 standard deviations above the mean.

Annual and growing-season total precipitation values are convenient for making comparisons, but to understand the dynamics of biotic activity one must consider the distribution of precipitation on a finer time scale (Table 2.4). Monthly precipitation was variable for all of the growing-season months. The 2 months of highest precipitation, May and June, also had the greatest year-to-year variation. Means and standard deviations for May and June precipitation in millimeters were

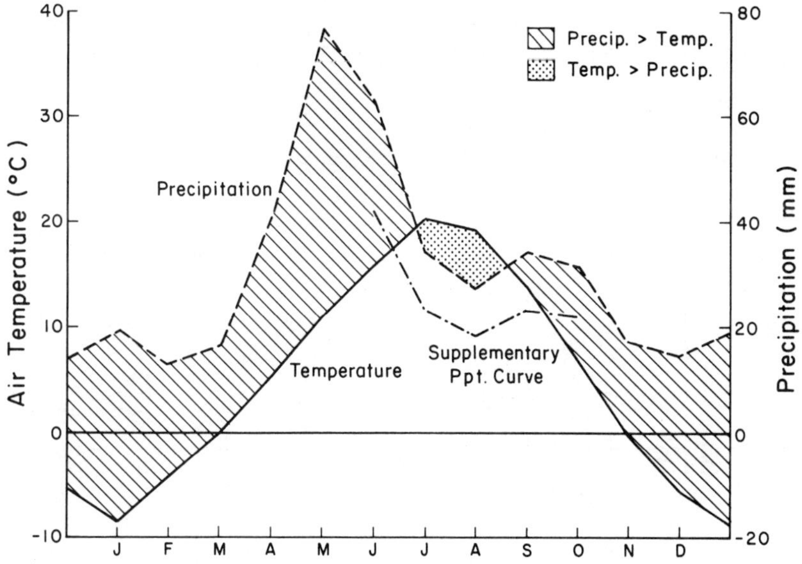

Figure 2.3. Climate diagram (Walter, 1979) for the experimental sites.

Table 2.4. Growing-Season Precipitation (mm) for the Experimental Sites for 1975–1979 Compared with Average at Sonnette, Montana

	Site I					Site II				
Date	1975	1976	1977	1978	1979	1976	1977	1978	1979	Sonnette (16 years)
May										
1–15	72	29	17	100	10	22	18	108	11	
16–31	32	68	28	164	41	22	28	174	41	
Total	104	97	45	264	51	44	46	282	52	77
June										
1–15	34	67	77	14	23	57	79	11	21	
16–30	75	35	14	43	21	37	19	46	18	
Total	109	102	91	57	44	94	98	57	39	63
July										
1–15	12	25	10	12	2	27	7	13	—	
16–31	17	13	8	17	41	14	7	17	42	
Total	29	38	18	29	43	41	14	30	42	34
August										
1–15	12	4	13	6	9	10	11	6	9	
16–31	2	6	5	6	8	7	7	5	10	
Total	14	10	18	12	17	17	18	11	19	27
September										
1–15	2	5	2	44	14	5	2	50	15	
16–30	4	9	34	9	1	9	38	11	1	
Total	6	14	36	53	15	14	40	61	16	34
Growing season	262	261	208	415	170	210	216	441	168	235

109 ± 96 and 78 ± 27. August precipitation was least variable, with a mean of 14 mm and a standard deviation of 3 mm.

Seasonal average air temperatures at approximately 2 m above the soil surface ranged from 16 to 18°C over the 4 years of measurement (Table 2.5). A large number of missing observations in the data from Site II in 1979 may explain the 2-degree difference between sites. Soil temperatures averaged 2 to 4 degrees greater than air temperature at the surface and very near to air temperature at 45 cm. Temperatures at a depth of 15 cm were slightly greater than air temperature. Wind speed was relatively constant, ranging from 2.2 to 2.7 $m \cdot s^{-1}$ over the 4 years.

As an example of the dynamics of these variables within a growing season, the data for Site I, 1979 (Table 2.6), is representative. Over the 7-month potential growing period, air temperatures ranged from 8°C in April, May, and October to above 20°C in July and August. Relative humidity was largely constant from April through August at approximately 50%. Soil temperatures followed air temperatures closely except for a lag in maximum temperature at 45 cm. Wind speed changed little during this period from 2.7 $m \cdot s^{-1}$ in April to 2.2 $m \cdot s^{-1}$ for the remainder of the months. Wind direction shifted from predominantly northerly and northwesterly in April and May to easterly and southeasterly in the remaining months.

As an index of the atmospheric demand for water at the sites, we present estimates of potential evapotranspiration (PET) calculated by the method of Van Bavel (1966) (Table 2.6). These values represent average PET because they were obtained using monthly averages of meteorological inputs. Critical here is that atmospheric demand for water always exceeds the supply and that maximum demand occurs in July and August.

2.2.3 Vegetation

2.2.3.1 Community Structure

The vegetation of the local region around the study sites is a *Pinus ponderosa*–grassland complex (Küchler, 1964; Payne, 1973; Ross and Hunter, 1976). Uplands are classified as northern mixed prairie (Brown, 1971) with open stands

Table 2.5. Growing Season Air Temperature (at 2 m), Soil Temperature, and Wind Speed (at 2 m) for Sites I and II, 1976–1979

	1976		1977		1978		1979	
	I	II	I	II	I	II	I	II
Air temperature (°C)	17	17	16	16	14	14	16	18
Soil temperature (°C)								
Surface	19	18	20	20	16	17	20	20
15 cm	18	18	21	—	16	14	18	19
45 cm	17	16	18	19	15	13	17	16
Wind speed ($m \cdot s^{-1}$)	2.7	2.7	2.7	2.7	2.2	2.7	2.2	2.2

Table 2.6. Growing-Season Air Temperature, Relative Humidity, Soil Temperature, Wind Speed, Wind Direction, and Potential Evapotranspiration (PET), Site I, 1979

	April	May	June	July	August	Sept.	Oct.
Air temperature (°C)	8	11	18	22	21	18	11
Relative humidity (%)	56	53	48	49	49	32	43
Soil temperature (°C)							
Surface	8	14	19	27	25	22	14
15 cm	7	12	18	25	23	20	14
45 cm	5	9	15	22	26	20	16
Wind speed ($m \cdot s^{-1}$)	2.7	2.2	2.2	2.2	2.2	2.2	2.2
Wind direction	N/NW	N/NW	E/SE	E	E/SE	E/SE	E/SE
PET[1] (mm)	70	110	120	165	155	125	60

[1]Calculated by the method of Van Bavel (1966) using average data.

of pine restricted to steep slopes and cups of ridges and buttes where soils are not well developed (Figure 2.4).

Vegetation of the study sites is representative of the northern mixed prairie in that it is dominated by cool-season mid-height grasses (Singh et al., 1983).

Figure 2.4. Typical northern mixed prairie in the vicinity of the study sites. *Pinus ponderosa* occupies slope and ridge topographic positions.

Table 2.7. Vascular Plants Encountered on Sites I and II during the 5 Years of the Experiment Along with Their Categorization into Functional Groups

	Cool Season	Warm Season
Grasses		
Agropyron smithii Rydb.	XXXXX[1]	
A. spicatum (Pursh) Scribn. & Smith	X	
Aristida longiseta Steud.		XXX
Bouteloua gracilis (H.B.K.) Griffiths		X
Bromus japonicus Thurb.	XXXX	
B. tectorum L.	XX	
Buchloe dactyloides (Nutt.) Engelm.		X
Calamagrostis montanensis (Scribn.) Scribn.	XX	
Carex filifolia Nutt.	X	
C. pensylvana Lan.	X	
Danthonia unispicata (Thurb.) Munro ex Mancova	X	
Festuca idahoensis Elmer	X	
Juncus interior Wieg.	X	
Koeleria cristata Pers.	XXXXX	
Muhlenbergia cuspidata (Torr.) Rydb.		X
Phleum pratense L.	X	
Poa pratensis L.	XXX	
P. sandbergii Vasey	XXXXX	
Schedonnardus paniculatus (Nutt.) Trel.		XXX
Sporobolus cryptandrus (Torr.) A. Gray		X
Stipa comata Tri. and Rupr.	XXX	
S. viridula Trinn.	XX	
Vulpia octoflora (Walt.) Rydb.	X	
Forbs		
Achillea millefolium var. *lanulosa* L.	XXX	
Agoseris glauca (Pursh.) D. Dietr.	X	
Allium textile A. Nels. and Macbr.	X	
Ambrosia psilostachya DC.		X
Androsace occidentalis Pursh.	XX	
Antennaria neglecta Greene		X
A. rosea Greene		X
Arabis holboellii Hornem.	X	
Arnica sororia Greene		
Artemisia ludoviciana Nutt.		X
Asclepias pumilla (A. Gray) Vail		X
Aster falcatus Lindl.		X
Astragalus crassicarpus Nutt.	XX	
A. drummondii Dougl.	X	
A. gilviflorus Sheldon	X	
A. purshii Dougl.	XX	
Bahia oppositifolia (Nutt.) DC.		XXX
Besseya wyomingensis (Nels.) Rydbg.	X	
Camelina microcarpa Andrz.	X	
Calochortus nuttallii Torr. and Gray	X	
Cerastium arvense L.	X	
Chenopodium album L.		X
Chorispora tenella (Pallas) DC.	X	
Cirsium undulatum (Nutt.) Spreng.		X

Table 2.7 *(continued)*

	Cool Season	Warm Season
Collomia linearis Nutt.	X	
Conyza canadensis (L.) Cronquist		X
Delphinium bicolor Nutt.	X	
Descurainia pinnata (Walt.) Britt.	X	
Echinacea pallida Nutt.		X
Erigeron divergens Torr. & Gray		XX
E. pumilus Nutt.		X
Erysimum asperum (Nutt.) DC.	X	
Geum triflorum Pursh.	X	
Grindelia squarrosa (Pursh.) Dunal		X
Haploppapus spinulosus (Pursh.) DC.		X
Hedeoma hispida Pursh.	XX	
Heterotheca villosa (Pursh.) Shinners		X
Lactuca serriola L.		X
Lappula echinata Gilib.	X	
Lepidium densiflorum Schrader	XXX	
Leucocrinum montanum Nutt.	X	
Lithospermum incisum Lehn.	X	
L. ruderale Dougl.	X	
Lomatium orientale Coult. and Rose	X	
Lupinus sericeus Pursh.	X	
Mammillaria missouriensis Sweet		Succulent[2]
Medicago sativa L.	X	
Melilotus alba Desr.	X	
M. officinalis (L.) Lanz.	XX	
Mertensia oblongifolia (Nutt.) G. Don.	X	
Monarda fistulosa L.	X	
Opuntia fragilis (Nutt.) Haw.		Succulent[2]
O. polyacantha Haw.		Succulent[2]
Orthocarpus luteus Nutt.		X
Oxytropis sericea Nutt.	X	
Penstemon nitidus Dougl.	X	
Dalea purpurea Vent.	X	
Phlox hoodii Rich.	XXX	
Plantago patagonica (Nutt.) Gray	X	
P. spinulosa DC.	XXX	
Polygonum viviparum L.		XX
Psoralea argophylla Pursh.	X	
P. esculenta Pursh.		X
Ranunculus glaberrimus Hook.	X	
Ratibida columnifera (Nutt.) Woot and Standl.		X
Sisyrinchium angustifolium Mill.	X	
Solidago missouriensis Nutt.		X
Sphaeralcea coccinea (Pursh.) Rydb.	XXX	
Taraxacum officinale Weber	XXXX	
Thlaspi arvense L.	X	
Tragopogon dubius Scop.	XXXX	
Vicia americana Muhl.	XX	
Viola nuttallii Pursh.	X	
Zygadenus venenosus S. Wats.	X	

Table 2.7 *(continued)*

	Cool Season	Warm Season
Half-shrubs and Shrubs		
Artemisia cana Pursh.	X	
A. dracunculus L.	X	
A. frigida Word.	XXX	
A. tridentata Nutt.	X	
Atriplex nuttallii S. Wats.	X	
Ceratoides lanata (Pursh.) J.T. Howell	X	
Gutierrezia sarothrae (Pursh.) Britt & Rusby	X	

[1]The number of Xs indicates relative importance based upon cover.
[2]Succulents were not given warm- or cool-season designations.

Agropyron smithii is the most frequent dominant grass throughout the northern mixed prairie as well as on our study sites. Unlike many locations in this grassland type, the warm-season shortgrasses, *Buchloe dactyloides* and *Bouteloua gracilis*, do not form a significant understory on the study sites.

Although more than 116 species of vascular plants were documented on the experimental sites during the 5-year investigation, only a few species in each of several functional groups were abundant. This, too, is characteristic of the northern mixed prairie and of most other grasslands. Plants of grasslands can be conveniently classified into six ecological or functional categories: warm-season grass, cool-season grass, warm-season forb, cool-season forb, shrubs, and succulents (Table 2.7). Although all categories were found on the experimental sites, warm-season grasses, warm-season forbs, succulents, and shrubs were low in abundance.

2.2.3.1.1 Canopy Cover

The general aspect of the grasslands in this region is not the dense sward often associated with grasslands. The basal portions of the plants occupy less than 50% of the ground area. At an oblique angle, one has the impression of a continuous cover of plants (Figure 2.5), but a steep vertical angle of observation one is struck by the sparseness of the cover.

Canopy cover at maximum development is quite variable from year to year (Figure 2.6). Total cover ranged from 85 to 110% on Site I and from 100 to 125% on Site II. Grasses accounted for 50% or more of total cover on both sites and *A. smithii* cover ranged from 20% on Site I to approximately 40% on Site II. Species composition on each site is illustrated with data from July 1977 (Table 2.8). The most important grasses on Site I were *A. smithii* and *K. cristata*. On Site II *A. smithii* and *Bromus japonicus* were the two grasses with greatest cover. *Achillea millefolium* and *Tragopogon dubius* were by far the most important forbs on Site I. On Site II *Taraxacum officinale* was the dominant forb. Cover of bare ground, lichens, litter, and rocks were essentially the same for the two sites.

Figure 2.5. Canopy cover of a northern mixed prairie site. View of a portion of replicate 2, high-SO_2 treatment, Site I.

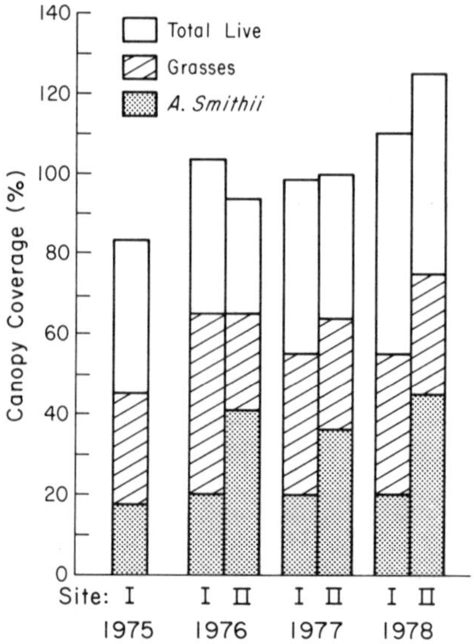

Figure 2.6. Canopy cover for total live vegetation, grasses, and *A. smithii* on Sites I and II over the 4 years of measurement.

Table 2.8. Canopy Cover by Species and Percentage Cover for Sites I and II July 1977

Species	Canopy Cover (%) Site I	Canopy Cover (%) Site II
Grasses		
Agropyron smithii	20	34
Aristida longiseta	2	<1
Bromus japonicus	4	13
Koeleria cristata	17	7
Poa pratensis	<1	1
P. sandbergii	6	8
Stipa comata	7	<1
Forbs		
Achillea millefolium	<1	2
Bahia oppositifolia	<1	2
Lepidium densiflorum	<1	5
Phlox hoodii	<1	4
Plantago spinulosa	<1	1
Sphaeralcea coccinea	2	<1
Taraxacum officinale	3	8
Tragopogon dubius	9	2
Half-shrubs		
Artemisia frigida	<1	1
Others		
Bare ground	5	8
Lichen	5	5
Moss	5	<1
Litter	74	80
Rock	<1	<1
Total grasses	56	65
Total forbs	25	26
Total shrubs	<1	1
Total vegetation	91	97

2.2.3.1.2 Biomass

The distribution of plant biomass in these grasslands is weighted toward belowground components (Figure 2.7). Belowground:aboveground ratios averaged 2.5 for both sites for the years of measurement. A generality that appears to have broad applicability to grasslands is that as one moves along a moisture gradient from wet to dry, belowground:aboveground ratios increase (Sims and Singh, 1978).

Aboveground biomass for both sites was dominated by cool-season perennial grasses (Dodd et al., 1982). Biomass of Site II contained a higher proportion of cool-season grasses (≅90%) than that of Site I (≅70%) and a correspondingly lower proportion of cool-season forbs (Table 2.9). *A. smithii* was the most important cool-season grass on both sites. Only slight variations in functional

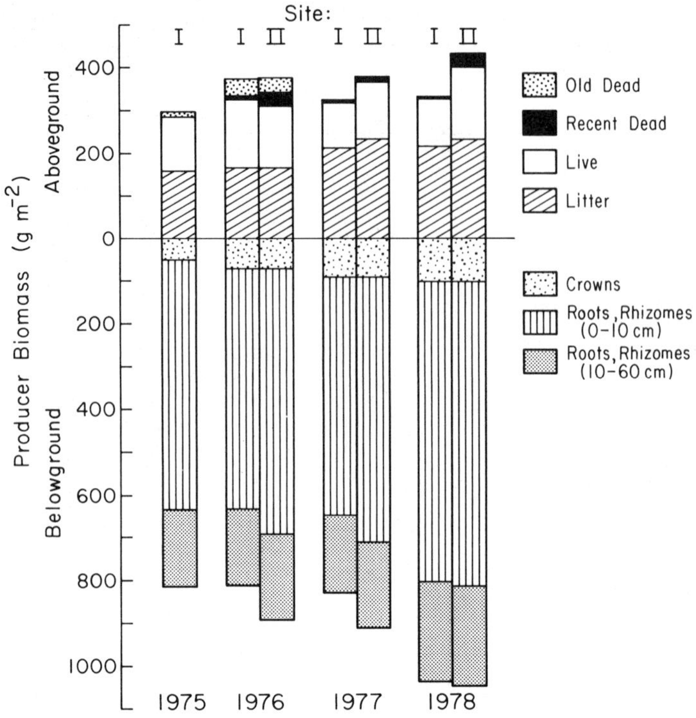

Figure 2.7. Distribution of biomass among vegetation components on Site I and Site II, 1975–1978, averaged across all treatments. Roots, rhizomes, crowns, and litter are on an ash-free oven-dried basis. Live, recent dead, and old dead are on oven-dried basis (Average ash content for aboveground parts was 7.75%).

Table 2.9. Standing Crop of Current Season's Aboveground Biomass ($g \cdot m^{-2}$) in July for Sites I and II

	Site I				Site II		
	1975	1976	1977	1978	1976	1977	1978
Cool-season grasses							
Agropyron smithii	56	69	39	48	91	92	119
Koeleria cristata	26	35	21	21	15	11	18
Others	20	16	16	16	46	23	51
Total cool-season grasses	102	120	76	85	152	126	188
Cool-season forbs	23	36	27	24	21	19	9
Other plants	9	10	7	7	3	5	5
Total standing crop	134	166	110	116	176	150	202

group contributions to total midseason biomass were noted in the 4 years of study.

Although cool-season forbs were a consistent component of the aboveground biomass on both sites, they were much less productive than the grasses and exhibited very little interannual variation. They were less important on Site II than on Site I. The sites, especially Site I, have a larger component of cool-season forbs than many other locations in the northern mixed prairie (Singh et al., 1983). Lauenroth and Whitman (1977), and Coupland (1973) reported cool-season forb contribution of less than 10% on sites in North Dakota, and Saskatchewan. Lauenroth et al. (1975) found cool-season forb contributions near 10% at other locations in Montana.

Collectively, warm-season grasses, warm-season forbs, half-shrubs, and succulents accounted for a greater proportion of aboveground biomass on Site I than on Site II (6 and 2%, respectively) but were low in abundance compared with other northern mixed-prairie sites (Singh et al., 1983).

Knowledge of the belowground components of grasslands is poor in relation to understandings of the aboveground component. Although roots of northern mixed-prairie plants may penetrate to a depth of 1 m or more, previous studies within this grassland type indicated that more than half of the belowground plant biomass was within the first 10 cm of the soil profile (Lauenroth and Whitman, 1977).

We limited most of our sampling of belowground biomass to the upper 10 cm, using total (live plus dead) plant biomass and its apportionment among roots, rhizomes, and crowns (basal parts of shoots that are subterranean but positioned above the transition zone of the plant). Reliable means of discrimination between species or species groups or between live and dead belowground material have not been developed, so further subdivisions were not possible. *A. smithii* was the only abundant species which possessed rhizomes, consequently we assumed that rhizome biomass could be attributed to *A. smithii.*

Average belowground biomass, based on monthly sampling from May to September, increased from 1975 to 1978 on Site I and from 1976 to 1978 on Site II (Table 2.10). This was most likely the result of exclusion of larger herbivores.

Table 2.10. Average Standing Crop ($g \cdot m^{-2}$ ash-free) of Belowground Biomass by Morphological Categories (0 to 10-cm Depth, Site I, 1975–1978, and Site II, 1976 and 1978)[1]

	Site I			Site II	
Component	1975	1976	1978	1976	1978
Crown	51	70	97	71	99
Rhizome	25	32	30	28	32
Roots	549	528	677	593	682
Total	625	630	804	692	813

[1]Comparable data were not collected in 1977.

Total belowground biomass in the upper 10 cm was similar on both sites and is comparable to that reported in other published data for the northern mixed prairie (Lauenroth and Whitman, 1977).

Approximately 85% of belowground biomass was in roots; 10% in crowns; and 5% in rhizomes. Morphological composition of belowground plant biomass was quite similar for the two sites and annual variation was small.

The structure of the vegetation at both of the experimental sites is similar using either canopy cover or biomass as the criterion. Cool-season grasses and specifically *Agropyron smithii* are clearly the dominants. Differences exist between the sites with respect to the contributions of subordinate species to either total cover or biomass. Additionally the absolute amount of biomass supported at Site II was greater than that for Site I.

2.2.3.2 Community Dynamics

2.2.3.2.1 Phenology

Growth initiation in these grasslands generally begins in late March, depending primarily upon the amount of snow cover. Earliest grasses are *Koeleria cristata* and *Poa secunda*, both cool-season grasses, while the two warm-season grasses, *Aristida longiseta* and *Bouteloua gracilis*, are the last grasses to initiate growth. The period of maximum vegetative growth is April and May or May and June for cool- and warm-season grasses, respectively. All cool-season grasses flower by early June except *A. smithii* which begins flowering in early July when both warm-season grasses flower. Generally, the cool-season grasses are dormant by early August although *A. smithii* remains green throughout the summer. By early September both *A. longiseta* and *B. gracilis* are dormant.

The pattern of the development of an individual tiller of *A. smithii* is illustrated using data for the development of leaves (Figure 2.8). Individual leaves are functional for only a portion of the growing season. During 1977 on Site I new leaves emerged each 12 days from early spring through late June. Leaves initiated early in the growing season had a large proportion of dead tissue by the time of peak standing crop of biomass (1 July). Leaf number 2, for example, initiated growth in early April, was 50% dead by mid-June, and almost entirely dead by late July. Losses to litter from current year's growth begins early in the growing season. The total leaf area of leaf number 2 showed a continual decline through the growing season after it had attained peak leaf area in mid-May.

Cool-season forbs are earlier to initiate flowering and to complete their annual life cycle than cool-season grasses. Two exceptions to this are *Sphaeralcea coccinea* and *Taraxacum officinale*. *S. coccinea* initiates growth approximately 4 weeks later than other cool-season forbs and lags approximately 4 to 6 weeks behind other species in initiating flowering, peak flowering, seed dispersal, and dormancy. *T. officinale* continues to grow and flower simultaneously throughout May and June, whereas other species seem to sequentially grow vegetatively and then flower.

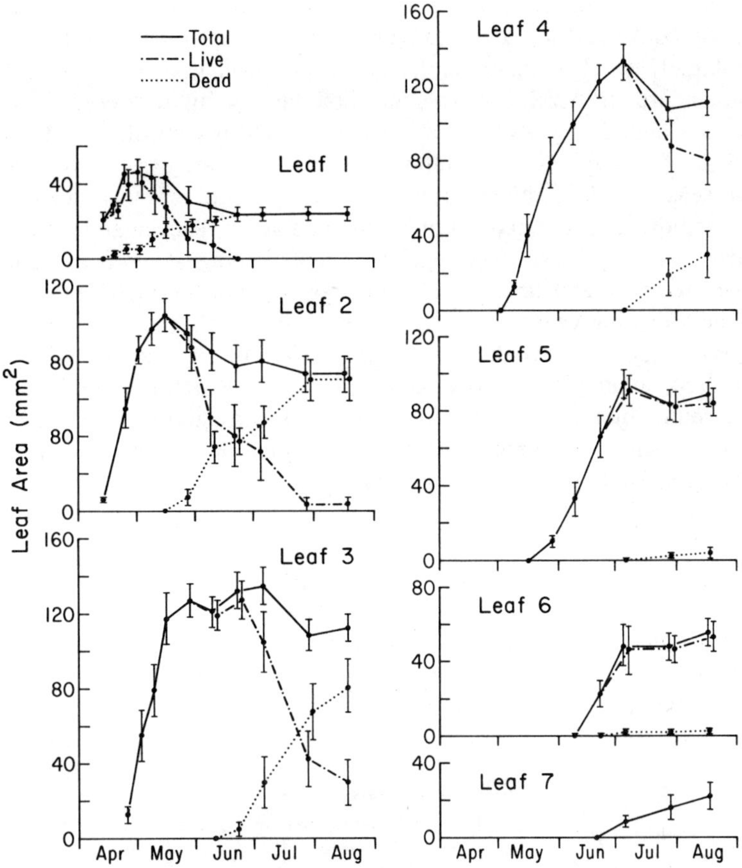

Figure 2.8. Pattern of leaf development in *Agropyron smithii* on Site I, control, 1979.

Warm-season forbs tend to initiate growth, flower, and complete their annual life cycles on dates similar to warm-season grasses. *Artemisia frigida*, a half-shrub, does not complete flowering until late September.

2.2.3.2.2 Biomass

The dynamics of plant biomass in northern mixed prairies is controlled by the strong seasonal nature of air temperatures favorable for plant growth and soil water amounts large enough to sustain growth. The period of favorable air temperatures is usually April through September. Soil water is usually high from April through June. Therefore, the period of maximum biomass production is usually April through June (Singh et al., 1983).

Our assessment of aboveground biomass included clipping the vegetation at the soil surface and separating the biomass of each species into three categories: live, current growing season's dead, and previous growing season's dead.

The typical seasonal pattern of aboveground biomass can be illustrated by an example of cool-season grasses (Figure 2.9). In April aboveground biomass consists largely of dead material from the previous growing season. Growth initiation occurs in April and is controlled largely by temperature. Biomass production is high during April, May, and June, and is controlled by the amount and timing of precipitation. Simultaneous with the increase in live biomass, previous season's dead biomass decreases. Current season's dead biomass increases rapidly after the maximum in live biomass is reached and is the product of the onset of a dry period which regularly occurs in August and September. After September, temperatures drop rapidly and snow may occur anytime through mid-May of the following year.

Analysis of biomass dynamics for several sites throughout the mixed-prairie region by Singh et al. (1983) indicated a high degree of similarity in the dynamics of aboveground biomass at our sites with sites throughout the mixed prairie. Productivity varied with precipitation amount and site fertility, but the general pattern is remarkably similar among sites.

2.2.4 Heterotrophs

The heterotrophic community was conceptualized as composed of three broad groups: vertebrates, invertebrates, and microflora. The first two groups are most often considered consumers while the last group is associated with decomposition.

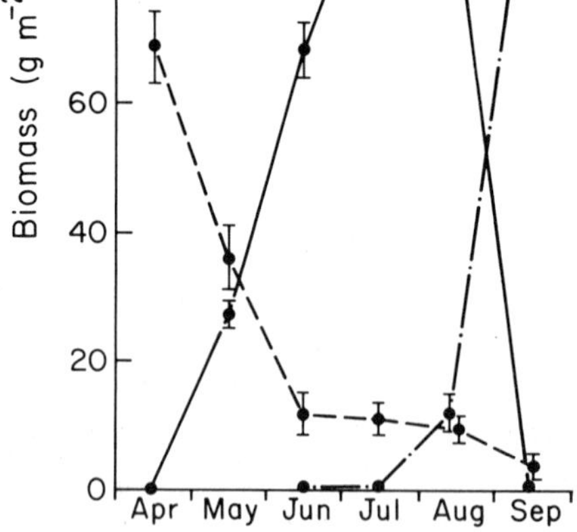

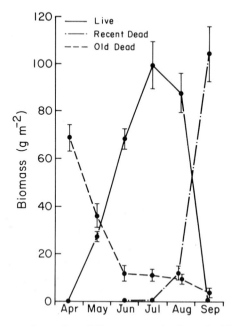

Figure 2.9. Growing-season dynamics of live, recent dead, and old dead aboveground biomass for cool-season grasses on Site I, 1975.

The distinction is not always so clear as this. Organisms associated with the first two groups and particularly the invertebrates are often thought of as decomposers or at least associated with decomposition processes.

Of the three groups of heterotrophs, we focused our attention on the invertebrates. Our experimental plots were too small to study the very mobile vertebrates adequately, and our objectives did not require that we allocate the large amount of resources required to study the microflora.

2.2.4.1 Invertebrate Community Structure

2.2.4.1.1 Composition of the Community

The major invertebrate taxonomic groups expected in grasslands are the arthropods, nematodes, tardigrades, rotifers, protozoans, and annelids. While neither the protozoans nor annelids were studied during this experiment, both are common to grassland soils (Wallwork, 1970). Soil nematodes are the most important invertebrate group in terms of nutrient and energy cycling. They may or may not contribute the greatest portion of soil invertebrate biomass, but their high metabolic rates and high densities account for a greater portion of nutrient and energy flow than any other group. The nematodes, along with the protozoans, rotifers, and tardigrades, are members of the soil water fauna; i.e., they live in the thin water films around soil particles (Wallwork, 1970). All four groups share the adaptive anabiosis characteristic, enabling them to survive times of severe soil water stress. The fact that all four groups are inhabitants of soil water films is of special concern in this study, as will be brought out later.

There are relatively few studies of the total arthropod communities of grasslands because these studies are so difficult to do effectively. The difficulties arise from the tremendous variation in size, form, habits, functional role, and developmental characteristics of the arthropods. There is currently no single technique for measurement by which the total arthropod fauna of a given locale can be accurately censused. Hence, during this study, the arthropod community was conceptually divided into three groups, each of which was censused by a different field technique. The techniques are discussed by Leetham et al. (1980a,b, 1982). The three groups were aboveground arthropods, soil macroarthropods, and soil microarthropods. Aboveground arthropods were those occurring in or above the soil surface litter and generally large enough to be collected in a sieve with 1.0-mm mesh. Soil macroarthropods were similar in size but occurred in the soil below the surface litter. Soil microarthropods occurred in and below the surface litter and were small enough to pass through the 1.0-mm-opening sieve.

The microarthropods were defined taxonomically as the soil Acarina, Pterygota (Insecta), Symphyla, and Pauropoda. The macroarthropods included the remainder of the insects and myriopods, which occur in the soil as developing immatures or as adults seeking refuge and/or food. Inevitably, overlap in total community censusing will occur with this system, but the overlap is small. All three arthropod-censusing efforts were coordinated spatially and temporally to achieve greatest accuracy in estimating the total arthropod community at given points in time. However, field censusing of tardigrades, rotifers, and nematodes was not coordinated in time or space with the arthropod census. While most of the arthropod work was done in the first several years of the studies, work on the other groups was done subsequently. Taxonomic identifications and biomass measurements generally were not made for these groups.

The arthropods identified from field sampling on the experimental sites in 1975 and 1976 include 80 insect families representing 17 orders, 67 mite (Acarina) families, 5 spider (Araneida) families, 4 tardigrade species, and a representative of Chilopoda (Geophilomorpha), Pauropoda, and Symphyla. Although voucher specimens of many species were sent to specialists for identification, an exhaustive species list could not be compiled. Many of the acarines and insect immatures have not been described. Therefore, the arthropod data were summarized and analyzed at the family or order level.

Tables 2.11 and 2.12 present a comprehensive overview of the arthropod community on the control plots of Sites I and II as estimated in 1976. Only the major arthropod groups, based on biomass, are included. Because all figures are seasonal means, the tables represent only the season as a whole. Overall, the soil microarthropods (predominantly acarina) are the dominant group in density, but rank a distant third in biomass. In terms of energy flow and total consumption per unit area, the microarthropods may rank near the top despite their small size because of their high metabolic rates. Although no statistical comparison of the two sites was made, Tables 2.11 and 2.12 indicate that between-site differences among the groups were small. The two sites appear to be reasonably good replicates in terms of the arthropod community size and taxonomic structure. We considered them representative of grasslands of the southeastern Montana area mainly because they closely resembled four sites, approximately 120 km to the northwest, that were sampled during the 1974 and 1975 seasons.

Density estimates for nematodes, rotifiers, and tardigrades could not be based on season means, since they were sampled on only one or two dates each season. Mid- to late-growing-season estimates showed nematodes reaching densities of $6 \times 10^6 \cdot m^{-2}$, rotifer densities of $3 \times 10^5 \cdot m^{-2}$, and tardigrade densities of $2 \times 10^4 \cdot m^{-2}$. These figures are based on counts and were not corrected for extraction efficiency. Active and inactive individuals were not distinguished, since the extraction technique used often quickly activates individuals from an anabiotic state. This is unfortunate because, for these organisms, total population size may not be as important as the percentage of active individuals.

It was noted earlier that most of the arthropod population in a grassland is associated with the soil or the surface litter. Table 2.13 indicates that, on the experimental sites, the biomass of soil-dwelling macro- and microarthropods is from 5 to 15 times greater than those aboveground. The predators show the highest ratio of below- to aboveground biomass. This is in part the result of the relatively large population of predatory beetles found in the soil as both larvae and adults; many of these forage aboveground nocturnally, but seek refuge in the soil during the day. All of the field sampling was accomplished in daylight.

While biomass estimates for the sites were not calculated for nematodes, rotifers or tardigrades, evidence from Willard (1974) and Scott et al. (1979) indicated

Table 2.11. Arthropod Community Structure on Site I in 1976[1]

Group (Order, Family)	Density (No. · m^{-2}) Above-ground	Density Soil Macros	Density Soil Micros	Biomass (mg · m^{-2}) Above-ground	Biomass Soil Macros	Biomass Soil Micros
Araneida	3			2		
Acarina (total)			107,500			44
Suborders:						
Mesostigmata			5,400			12
Prostigmata			84,900			19
Cryptostigmata			16,800			12
Astigmata			400			<1
Diplura			200			2
Collembola	174		24,400	4		24
Orthoptera						
Acrididae	<1	1[2]		2	<1	
Thysanoptera						
Thripidae	83			3		
Hemiptera (total)	12			10		
Cydhidae	<1			2		
Lygaeidae	9			3		
Micridae	1			1		
Nabidae	<1			1		
Scutelleridae	<1			1		
Homoptera (total)	50	11		14	19	
Ceropodiae	<1			<1		
Cicadellidae	13			10		
Pseudococcidae	15	11		1	19	
Coleoptera (total)	56	114		64	508	
Carabidae	5	43		8	319	
Curculionidae	10	24		29	71	
Elateridae	<1	16		<1	96	
Staphylinidae	7			4		
Tenebrionidae	<1			1		
Chrysomelidae	10	29		12	11	
Cicindelidae		1			12	
Mecoptera						
Borneidae		5			5	
Lepidoptera	1	46		6	537	
Noctuidae	<1			3		
Pyraldae		45			536	
Diptera (total)		19			56	
Asilidae		8			42	
Hymenoptera	23			15	227	
Formicidae	21	153		14	227	
Total arthropods[3]	420	350	132,700	121	1,353	70

[1]Values are seasonal mean densities and biomass based on six sample dates.
[2]Eggs.
[3]Totals are not limited to groups listed but include all groups censused.

Table 2.12. Arthropod Community Structure on Site II in 1976[1]

Group (Order, Family)	Density (No. · m^{-2}) Aboveground	Density Soil Macros	Density Soil Micros	Biomass (mg · m^{-2}) Aboveground	Biomass Soil Macros	Biomass Soil Micros
Araneida	3			3		
Acarina (total)			106,500			40
Suborders:						
Mesostigmata			4,400			8
Prostigmata			85,700			19
Cryptostigmata			14,900			13
Astigmata			1,500			1
Diplura			200			2
Collembola			19,200			17
Orthoptera						
Acrididae	<1			2		
Thysanoptera						
Thripidae	31			1		
Hemiptera (total)	10			10		
Cydhidae	<1			4		
Lygaeidae	8			3		
Homoptera (total)	20	30		6	44	
Cicadellidae	4			4		
Pseudococcidae	12	11		1	44	
Coleoptera (total)	49	149		80	1,256	
Carabidae	4	103		8	1,053	
Chrysomelidae	9	3		15	11	
Curculionidae	16	33		48	108	
Elateridae	<1	8		2	79	
Orthoperidae	5			1		
Staphylinidae	6			3		
Tenebrionidae	<1			3	5	
Mecoptera						
Borneidae		10			11	
Lepidoptera	1	18		3	211	
Noctuidae	<1			1		
Pyralidae		18			211	
Diptera (total)		18			85	
Asilidae		16			83	
Hymenoptera	21	52		10	23	
Formicidae	20	52		10	23	
Total arthropods[2]	158	275	126,700	117	1,631	59

[1]Values are seasonal mean densities and biomass based on six sample dates.
[2]Totals are not limited to groups listed but include all groups censused.

that, if all invertebrates were considered, the importance of the aboveground portion of the population would be even further diminished. This is not surprising, considering the high proportion of primary production retained as plant crowns and roots, especially in combination with the amelioration within the soil of the harsh and relatively xeric surface conditions characteristic of semiarid grasslands.

The trophic structure of the arthropod community was estimated by categorizing each of the arthropods collected by trophic or functional status. Trophic

Table 2.13. Aboveground:Belowground Ratios Based on Seasonal Mean Biomass[1]

Group	Site I 1975	Site I 1976	Site II 1976
Total arthropods	6	13	15
Herbivores[2]	4	8	12
Predators	24	32	58
Omnivores	7	16	2
Scavengers	1	11	5
Fungivores	+[3]	+[3]	+[3]
Nonfeeding	19	1100	+[3]

[1]See text for explanation of terms aboveground and belowground.
[2]Includes all anthropods that feed on plants.
[3]+100% occurred belowground.

assignments were based on published information, gut content analyses, and information from specialists. Herbivores were the dominant group in both the aboveground arthropods and soil macroarthropods, but ranked third behind fungivores and predators in the soil microarthropods. The proportion of predators was high in relation to herbivores in the soil macroarthropods; however, many of the predators categorized as soil macroarthropods actually function as nocturnal, aboveground predators. Therefore, the soil secondary predators are supported by a greater primary consumer biomass than initially appears. Also, many of the first instar larvae of the predatory insects probably feed on nematodes whose biomass was not included in the nonpredator category. Relative turnover rates are another consideration in the predatory–nonpredator relationship. Many of the herbivorous arthropods are probably multivoltine, while most predatory types are probably univoltine. Herbivory was much reduced in the microarthropods, a group that utilizes the tremendous fungal, nematode, and protozoan resources in the soil.

A general comparison of season mean nonpredatory:predator biomass ratios on the control plots shows a high degree of similarity between Sites I and II in 1976, but a striking dissimilarity between 1975 and 1976 on Site I (Table 2.14). On Site I, predator biomass appeared much reduced in 1975, whereas the nonpredator biomass was about the same in both years. The ratios for both sites in 1976 are comparable to those estimated for a South Dakota mixed-prairie site and a shortgrass steppe site in northeastern Colorado (French et al., 1979). French et al. (1979) calculated nonpredator:predator ratios of 1.5 and 3.6 for the South Dakota and Colorado sites.

2.2.4.1.2 Invertebrate Spatial Distribution

The most difficult problem confronting invertebrate censusing is extreme sample variability resulting from patchy or clumped distributions. For example, at the economically and physically feasible level of sampling done during this study, sample variability was so high that standard errors of the means for most taxonomic and trophic groups on all dates were greater than 25% of the mean and

Table 2.14. Nonpredator:Predator Ratios for Sites I and II Control Plots for 1975 and 1976[1]

Site and Year	Aboveground Arthropods	Soil Macroarthropods	Soil Microarthropods	Aboveground and Soil Macroarthropods	All Groups Combined
Site I, 1975	29	7	2	8	7
Site I, 1976	8	3	3	3	3
Site II, 1976	10	2	3	2	2

[1]Figures based on mean seasonal biomass.

were often as high as 100% of the mean. Estimates within 20% of the mean were often reached only when groups were combined or total densities and biomass were used.

Clumped or patchy arthropod distributions are unquestionably the result of microhabitat variability. Distribution of many herbivorous arthropods, and associated predators, can be related directly to the distribution of their host plants. For example, population estimates for a small scorpion fly, *Boreus nix* Carp. (Mecoptera:Boreida), in 1975 suggested substantially different densities on the two adjacent replicates of the Site I control plot. Scorpion flies are closely associated with mosses (Borrer and DeLong, 1971), and vegetation sampling during the same season showed that a moss (*Polytrichum piliferum*) occurred in much higher densities on the replicate where the scorpion flies were also higher. Similar associations between insects and hosts or other habitat requirements probably existed on the field plots but were not resolved during our studies. Microhabitat variability in the soil and litter is not always readily observable or measureable; nevertheless, it must be assumed to occur because seemingly identical habitats can contain drastically different densities of the same invertebrate species. In this study, for example, tardigrade densities per sample ranged from 0 to over 300 in soil cores taken from seemingly homogeneous stands of *Agropyron smithii.* Such ranges were also common for microarthropods and nematodes.

Vertical distribution patterns of the soil-dwelling invertebrates vary with invertebrate group and with species within a group. For example, tardigrades were almost completely restricted to near the soil surface; 97% of the numbers were found in the top 2 cm (Leetham et al., 1982). However, one species was restricted to the thin surface litter zone, while others were more commonly found just below the soil surface. Nematodes were distributed deeper in the soil profile than the tardigrades, but a majority were found in the top 10 cm (Leetham et al., 1982). Anderson (1978) found that over 80% of the nematodes in two soil types in a northeastern Colorado shortgrass steppe occured in the top 20 cm of the profile, and most were in the top 5 cm. By contrast, the predatory Monochida group had significantly higher densities in the deeper soil layers (below 10 cm).

Vertical distribution patterns of some soil microarthropods are distinctly different from other invertebrates. Although no samples were taken on the experimental plots for this purpose, samples from a nearly similar grassland and extensive sampling from a northeastern Colorado shortgrass steppe showed that microarthropods occur in high densities at substantial depths (Figure 2.10). On the other hand, over 90% of rotifers were found in the 0- to 10-cm soil layer on the experimental sites. We can conclude that the soil microarthropod community is composed of both predominantly surface-dwelling and predominantly subsurface-dwelling portions, each adapted to the prevailing conditions. The shortgrass steppe data indicate that no significant movements of either group into the area of the other occurred diurnally or seasonally. The deep-dwelling microarthropods appeared to be distributed in relation to soil water conditions, peaking where the most consistently moist soil layers occurred (at about 30 cm).

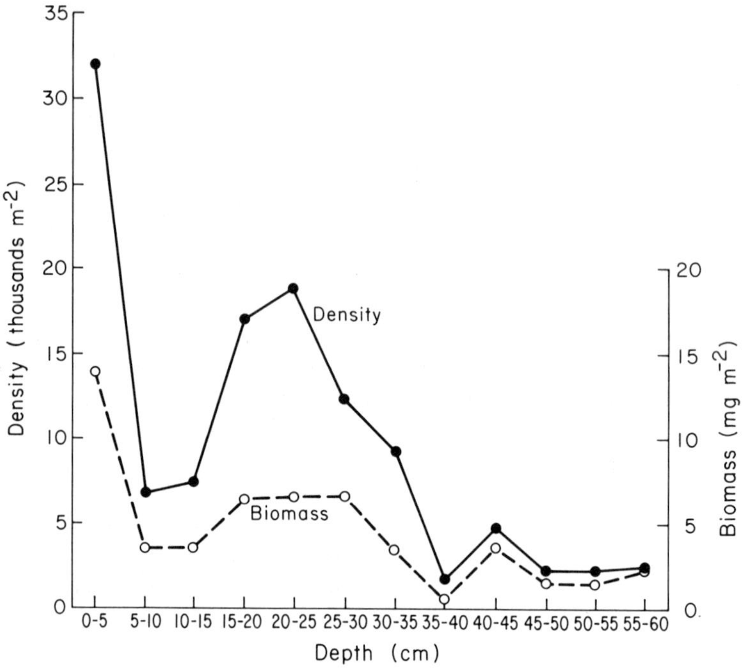

Figure 2.10. Vertical distribution of soil microarthropods in the top 60 cm of the soil of a southeastern Montana grassland. Density (thousands · m^{-2}), biomass (mg · m^{-2}).

Soil macroarthropods were largely restricted to the top 15 cm of the soil profile. One notable exception was cicadid (Homoptera:Cicadidae) nymphs, which were commonly found at 40–60 cm when deep soil pits were dug to describe the soil. It is apparent, however, that the large preponderance of the soil invertebrate biomass is distributed in relation to plant root biomass, which is concentrated near the soil surface, and decreases rapidly with depth below about 20 cm.

2.2.4.2 Invertebrate Community Dynamics

Biomass dynamics of the arthropod community was variable as a function of both location and group (Figure 2.11). It was expected that a seasonal peak for the aboveground arthropods would occur at or shortly after the time of peak herbage standing crop, since herbivores contributed the overwhelming majority of aboveground biomass. However, consistent seasonal trends did not appear for the aboveground arthropods for any of the 3 years studied. While a collection of adult grasshoppers (*Melanoplus* sp.) is known to have caused the late-summer 1975 biomass peak on Site I, density peaks were similarly erratic. It must be remembered that this aboveground biomass represents a small proportion of the total for the arthropod community and that some aboveground consumption is done by arthropods categorized as soil dwellers. Further, the aboveground population includes mostly adult insects, and the curves may represent population

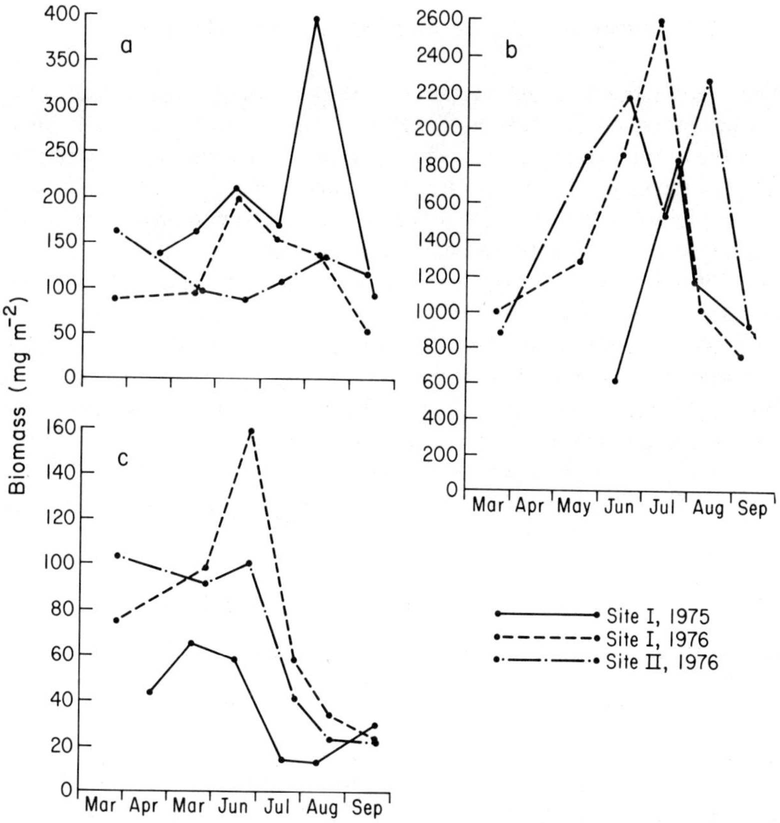

Figure 2.11. Seasonal patterns of arthropod biomass (mg·m^{-2}) Sites I and II, 1975 and 1976. (a) aboveground, (b) soil macroarthropods, (c) soil microarthropods.

peaks for a variety of species with different growth patterns and vegetation preferences that the data was insufficiently detailed to separate.

Distinct seasonal biomass peaks were noted for both soil macro- and microarthropods in all 3 site-years (Figure 2.11). The peak for both the herbivorous and predatory macroarthropods did occur, as expected, near the time of peak standing herbage. Since this population is made up largely of immatures, this parallel to plant growth is reflected in both density and individual development for many insect species. On the other hand, the soil microarthropods are dependent on fungal and nematode biomass for food, and hence are controlled more by soil water dynamics than by plant growth trends. The seasonal decline in soil microarthropod biomass (Figure 2.11) coincides with the decline in soil moisture after the spring and early summer rainy season (see Figure 2.2, Section 2.3).

While generalities about the heterotrophic community are difficult, we have no reason to believe that differences between the experimental sites were out of proportion to those discussed for soils or vegetation.

2.3 Summary of Site Similarities and Differences

Our objectives for this chapter were to describe the experimental sites, compare them to each other, and discuss how well they represented the grasslands of the northern mixed-prairie region. We used two criteria to limit the descriptions. The first related to the general setting of the experiment. We included only those items that we considered critical to understanding the experimental environment. The second criterion considered those characteristics that we would rely upon in our discussions of the response of the grasslands to SO_2 exposure.

Table 2.15 summarizes several of the key features of the sites and illuminates the important similarities and differences. Site I has deeper soils with less clay and organic matter than Site II. Site II supports a slightly greater canopy cover of vegetation and considerably more biomass. Both sites contain cool-season grasses and forbs as the dominants in terms of canopy cover and biomass. *Agropyron smithii* is the dominant species on both sites, comprising greater than 20% of the canopy cover of vegetation (Table 2.8) and 40 to 60% of the aboveground biomass

Table 2.15. Summary of Important Similarities and Differences between the Experimental Sites

Characteristic	Site I	Site II
Soils		
Depth	>134 cm	>100 cm
Texture[1]	Loam	Clay loam
Organic matter[1]	1%	3%
Vegetation		
Canopy cover	91%	97%
Cool-season grasses	54%	64%
Forbs	25%	26%
Biomass	$132\ g \cdot m^{-2}$	$176\ g \cdot m^{-2}$
Cool season grasses[2]	$95\ g \cdot m^{-2}$	$155\ g \cdot m^{-2}$
Belowground[3]	$685\ g \cdot m^{-2}$	$752\ g \cdot m^{-2}$
Heterotrophs		
Density		
Aboveground anthropods	$420 \cdot m^{-2}$	$158 \cdot m^{-2}$
Soil macroarthropods	$350 \cdot m^{-2}$	$275 \cdot m^{-2}$
Soil microarthropods	$132{,}700 \cdot m^{-2}$	$126{,}700 \cdot m^{-2}$
Biomass		
Aboveground arthropods	$121\ mg \cdot m^{-2}$	$117\ mg \cdot m^{-2}$
Soil macroarthropods	$1{,}353\ mg \cdot m^{-2}$	$1{,}631\ mg \cdot m^{-2}$
Soil microarthropods	$70\ mg \cdot m^{-2}$	$59\ mg \cdot m^{-2}$
Nonpredator:predator ratios (1976)		
Aboveground arthropods	8	10
Soil macroarthropods	3	2
Soil microarthropods	3	3

[1]A horizon.
[2]Averages across years in July standing crop.
[3]0–10 cm.

(Table 2.9). Aboveground:belowground biomass ratios were approximately 0.2 for both sites.

Density of arthropods was greater at Site I than Site II, particularly aboveground (Tables 2.1 and 2.12). Biomass of arthropods was more similar than density, as were nonpredator:predator ratios.

We believe that, within the scope of the objectives, our sites are reasonable experimental replicates (Table 2.15). To a very large extent we are interested in generalities about the nature of the response of these mixed-prairie sites to SO_2 exposure with the intent to make statements about the potential responses of grasslands across the mixed-prairie region.

Based upon published data for sites in the northern mixed prairie (Singh et al., 1983), we consider the vegetation on our sites to fall well within the northern mixed-prairie type. Singh et al. (1983) reported ranges for aboveground biomass (live + dead) (90–472 $g \cdot m^{-2}$), aboveground biomass (live) (56–139 $g \cdot m^{-2}$), and litter biomass (136–393 $g \cdot m^{-2}$). Corresponding values for our Sites I and II are 140 and 159 $g \cdot m^{-2}$ for aboveground biomass (live + dead), 74 and 71 $g \cdot m^{-2}$ for aboveground biomass (live), and 153 and 136 $g \cdot m^{-2}$ for litter biomass. The average contributions of warm- and cool-season functional groups to aboveground net primary production for the region was: cool-season grasses, 70%; warm-season grasses, 10%; cool-season forbs, 8%; and warm-season forbs, 3%. Similar values for Sites I and II were: cool-season grasses, 75 and 88%; warm-season grasses, 2 and <1%; cool-season forbs, 18 and 10%; warm-season forbs, 1 and <1%, Aboveground net primary production reported by Singh et al. (1983) for the northern mixed-prairie region ranged from 51 to 679 $g \cdot m^{-2} \cdot year^{-1}$ with most sites between 150 and 250 $g \cdot m^{-2} \cdot year^{-1}$. Aboveground net primary production for our sites is given in Table 5.7, Section 5.3.4. Values for Site I ranged from 120 to 200 $g \cdot m^{-2} \cdot year^{-1}$ and for Site II from 140 to 270 $g \cdot m^{-2} \cdot year^{-1}$. Unfortunately, no similar data set for comparing heterotrophs is available.

References

Aandahl, A. 1972. *Soil Map of the Great Plains.* U.S. Soil Conservation Service, Lincoln, Nebraska.

Anderson, R. V. 1978. *Free-Living Nematode Population Dynamics: Effects on Nutrient Cycling.* Ph.D. Thesis. Fort Collins, Colorado: Colorado State Univ.

Borrer, D. J., and D. M. DeLong. 1971. *An Introduction to the Study of Insects.* 3d Ed. New York: Holt, Rinehart, and Winston.

Brown, R. W. 1971. Distribution of plant communities in Southeastern Montana badlands. *Am. Midl. Natur.* 85:458–477.

Coleman, D. C., R. Andrews, J. E. Ellis and J. S. Singh. 1976. Energy flow and partitioning in selected man-managed and natural ecosystems. *Agro-Ecosystems* 3:45–54.

Coupland, R. T. 1973. Producers: I. Dynamics of aboveground standing crop. *Canadian IBP Matador Project Tech. Rep. No. 27.* Saskatoon: Univ. Saskatchewan.

Coupland, R. T. 1979. The nature of grassland. In R. T. Coupland, ed. *Grassland Ecosystems of the World*, pp. 23–30. Cambridge: Cambridge Univ. Press.

Dodd, J. L., W. K. Lauenroth, and R. K. Heitschmidt. 1982. Effects of controlled SO_2 exposure on net primary production and plant biomass dynamics. *J. Range Manage.* 35:572–579.

Fenneman, N. M. 1931. *Physiography of the Western United States.* New York: McGraw-Hill.

French, N. R., R. K. Steinhorst, and D. M. Swift. 1979. Grassland biomass trophic pyramids. In *Perspectives in Grassland Ecology*, N. R. French, ed. pp. 59–88. New York: Springer-Verlag.

Küchler, A. W. 1964. *Potential Natural Vegetation of the Coterminous United States.* New York: American Geographical Society.

Lauenroth, W. K., J. L. Dodd, R. K. Heitschmidt, and R. G. Woodmansee. 1975. Biomass dynamics and primary production in mixed prairie grasslands in Southeastern Montana: Baseline data for air pollution studies. pp. 559–578. Fort Union Coal Field Symposium, Billings: Eastern Montana College.

Lauenroth, W. K., and W. C. Whitman. 1977. Dynamics of dry matter production in a mixed-grass prairie in western North Dakota. *Oecologia (Berl.)* 27:339–351.

Leetham, J. W., T. J. McNary, J. L. Dodd, and W. K. Lauenroth. 1982. Response of soil nematodes, rotifers and tardigrades to three levels of season-long sulfur dioxide exposure. *Water Air Soil Pollut.* 17:343–356.

Leetham, J. W., J. L. Dodd, R. D. Deblinger, and W. K. Lauenroth. 1980a. Arthropod populations responses to three levels of chronic sulfur dioxide exposure in a northern mixed-grass ecosystem. I. Soil microarthropods. pp. 139–157. In *The Bioenvironmental Impact of a Coal-Fired Power Plant*, E. M. Preston, D. W. O'Guinn, and R. A. Wilson, eds. Sixth Interim Report. Colstrip, Montana, EPA-600/3-81-007, Corvallis, Oregon: Environmental Protection Agency.

Leetham, J. W., J. L. Dodd, R. D. Deblinger, and W. K. Lauenroth. 1980b. Arthropod population responses to three levels of chronic sulfur dioxide exposure in a northern mixed-grass ecosystem. II. Aboveground arthropods. pp. 158–175. In *The Bioenvironmental Impact of a Coal-Fired Power Plant*, E. M. Preston, D. W. O'Guinn, and R. A. Wilson, eds. Sixth Interim Report. Colstrip, Montana, EPA-600/3-81-007, Corvallis, Oregon: Environmental Protection Agency.

Mack, R. N. and J. N. Thompson. 1982. Evolution in steppe with few large hooved mammals. *Am. Naturl.* 119:757–773.

Northern Great Plains Resources Program. 1975. *Effects of Coal Development in the Northern Great Plains.* Billings, MT: Staff Report.

Paul, E. A., F. E. Clark, and V. O. Biederbeck. 1979. Microorganisms. In *Grassland Ecosystems of the World: Analysis of Grasslands and Their Uses*, pp. 87–96. R. T. Coupland, ed. Cambridge: Cambridge Univ. Press.

Payne, G. F. 1973. Vegetative rangeland types in Montana. *Montana Agric. Exp. Sta. Bull. 671* Bozeman: Montana State Univ.

Ross, R. L., and H. E. Hunter. 1976. Climax Vegetation of Montana Based on Soils and Climate. U.S. Dep. Agric., Soil Conserv. Serv.

Scott, J. A., N. R. French, and J. W. Leetham. 1979. Patterns of consumption in grasslands. In *Perspectives in Grassland Ecology*, N. R. French, ed. pp. 89–105. New York: Springer-Verlag.

Sims, P. L., and J. S. Singh. 1978. The structure and function of ten western North American grasslands. III. Net production, turnover, and efficiencies of energy capture and water use. *J. Ecol.* 66:573–597.

Singh, J. S., W. K. Lauenroth, R. K. Heitschmidt, and J. L. Dodd. 1983. Structural and functional attributes of the vegetation of the northern mixed prairie of North America. *Botanical Rev.* 49:117–149.

Van Bavel, C. H. M. 1966. Potential evaporation: The combination concept and its experimental verification. *Water Resources Res.* 2:455–467.

Walter, H. 1979. *Vegetation of the Earth and Ecological Systems of the Geo-biosphere.* 2nd Ed. New York: Springer-Verlag.

Wallwork, J. A. 1970. *Ecology of Soil Animals*. London: McGraw–Hill.
Willard, J. R. 1974. Soil invertebrates: VIII. A summary of populations and biomass. *Canadian/IBP Grassland Biome Tech. Rep. No. 56*. Saskatoon, Saskatchewan, Canada: Univ. of Saskatchewan.

3. The Field Exposure System

ERIC M. PRESTON AND JEFFREY J. LEE

3.1 Introduction

The dynamics of SO_2 exposure concentrations at any point are unique, determined by distance from various sources, source strengths, wind speed and direction, topography, stack height, and a variety of lesser influences. Nevertheless, in the simplest analysis, distinction can be made between exposure patterns in the vicinity of point (single) sources and those influenced by area (multiple) sources.

Concentrations of SO_2 near a point source with tall stacks such as a modern power plant are normally background. Concentrations above background are rare. Area sources, resulting from multiple emission sources, tend to cause higher background concentration levels and lower variability in concentrations than for point sources.

We decided to study the response of a mixed prairie grassland to exposure frequencies typical of area sources. The rationale for this is based both in the perceived applicability of the results to the northern Great Plains and in the desire to maximize the potential to develop new understanding of the action of SO_2 at the system level. In 1973, vulnerability of the U.S. energy supply to vagaries of the import market was acutely perceived. Plans for rapidly developing western coal reserves were materializing and it seemed likely that within the succeeding decades many coal-fired power plants and other coal conversion facilities would be developed (Durran et al., 1979; Northern Great Plains Resources Program, 1975;

White et al., 1979). It appeared that the northern Great Plains grasslands would be exposed to SO_2 from multiple sources having overlapping plume paths. In addition, most previous ecological field work on the effects of SO_2 had related to effects from acute exposures associated with point sources. These results suggested that xeric-adapted vegetation was less susceptible to acute injury than vegetation from habitats having high relative humidity. Therefore, the likelihood of acute effects appeared to be minimal. On the other hand, low-level exposure appeared likely and little knowledge existed to evaluate the potential for such effects.

Existing exposure systems such as open top chambers (Heagle et al., 1973) were not suitable for the present study, since they require more power than could be provided to the remote grassland site and they are vulnerable to damage by high winds. If the responses of the grassland were to be studied, a system had to be devised which could expose plots over several growing seasons to SO_2 concentrations having temporal dynamics similar to those expected to result from anthropogenic emissions. Disturbance of the microclimate and biota had to be minimal, and the area exposed on the spatial scale of the populations to be sampled had to be large to reduce edge effects and assure adequate population and sample sizes.

In area sources, concentrations are often highly correlated over short time intervals, leading to a lognormal distribution of concentration frequencies (Gifford, 1974; Knox and Pollack, 1974). In practice, concentration frequency curves of area sources can be described by a lognormal distribution over a broad range of pollutant types and source configurations (Larsen, 1971; Pollack, 1975). Therefore, the temporal sequence of pollutant concentrations on the experimental plots should approximate lognormal distributions. The median exposure treatment concentrations must cover the range of median concentrations measured near area sources. The system described below generally met these criteria.

3.2 The Zonal Air Pollution System (ZAPS)

Sulfur dioxide exposure of plots using this system was initiated at Site I in May 1975. A second set of plots (Site II) began in April of 1976. The experimental treatments lasted for five growing seasons on Site I and for four seasons on Site II. Sulfur dioxide exposures were continuous during the thermal potential growing season (early April through mid-October). Since the dominant plants on the sites can be photosynthetically active at temperatures as low as 5°C (Singh et al., 1980), we defined the growing season as the period when the 10-day running average temperature exceeded 5°C. This ensured that SO_2 exposure was neither begun nor stopped in response to brief periods of unseasonal weather. The spatial relationships of the two field sites and the visual impact of one growing season of SO_2 treatment can be seen in Figure 3.1.

Figure 3.1. Layout of experimental sites. Each site has a control and three treatment plots receiving successively greater median SO_2 concentrations. On both sites, the control plot is on the extreme left and the highest treatment plot on the extreme right. Scale is 1:12,500. Photographed June 1976, John Taylor.

3.2.1 The Gas Delivery System

Each gas delivery system consisted of a network of 2.5-cm-inside-diameter (ID) aluminum pipes set parallel to and approximately 0.7 m above the ground and supported at 6.1-m intervals by pipes driven 0.8 m into the ground (Figure 3.2) (Preston and Lee, 1982). Sulfur dioxide release points (0.8-mm horizontal holes) were placed at 3-m intervals so that no location within a plot was more than 5.5 m from a source. A continuous flow of air through the lines was maintained by a helical compressor.

Each study site consisted of four 0.5-ha treatment plots located along a line with 61-m buffer zones to reduce interference between plots. Livestock were excluded

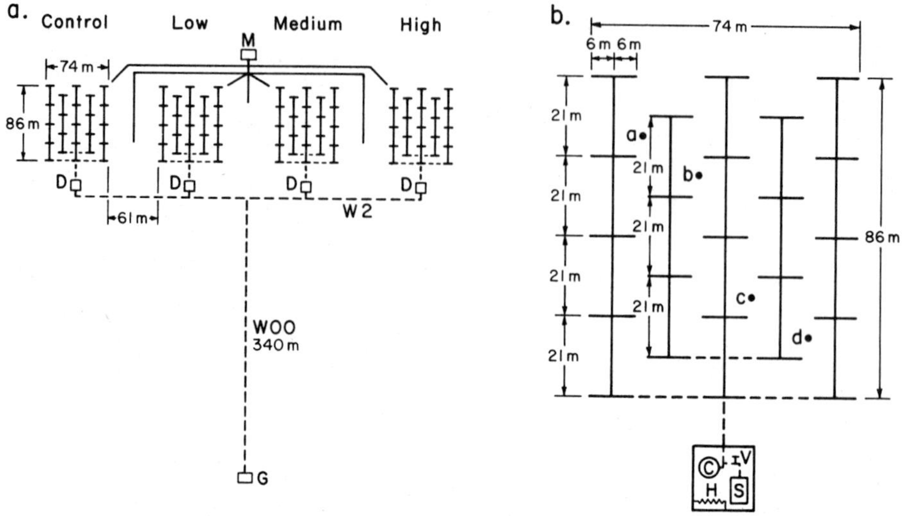

Figure 3.2. (a) Schematic of the Zonal Air Pollution System showing the individual delivery and the common monitoring and electrical systems. Control, low, medium, and high indicate SO_2 exposures on the experimental plots. (b) An expanded view of a single plot. Key: a–d—real time monitoring stations; C—compressor; D—delivery stations; G—6-kW Diesel generator; H—1-kW heater; M—central monitoring station; S—sulfur dioxide tanks; V—valve; WOO—three conductor o/o copper wire; W2—three conductor two aluminum wire.

to protect them from injury and to protect equipment from damage. On all but one of the four plots, SO_2 was bled into the airstream of the pipe network at a constant rate. The fourth plot received no direct input of SO_2, but did receive some SO_2 drift from other plots (Figures 3.1, 3.2).

Sulfur dioxide was released to the three treatment plots at rates that were roughly proportional to the desired concentrations. By convention, these treatments were designated low, medium, and high. The plot receiving no direct SO_2 was designated control. Concentrations on the plots varied as changing meterological conditions caused different degrees of dilution and deposition.

By utilizing many small, elevated, dilute point sources (over 250 per 0.5 ha), adequate dilution of SO_2 at ground level was ensured and, in effect, an area source was created. This prevented step-function changes in concentrations in time and space ("hot spots"), except at short distances (1–2 m) from a source. These hot spot areas were excluded from sampling.

3.2.2 The Sulfur Dioxide Monitoring System

Sulfur dioxide concentrations on the plots were monitored by recording the output from a Meloy Model SA160-2 flame photometric sulfur gas analyzer.

Ambient air samples were continuously drawn through eight Teflon sample lines by a time-sharing device. One of the lines passed through a charcoal filter, providing a check for instrument drift. The outer seven lines extended to monitoring locations on the plots, with at least one per plot. Each sample line was monitored for 7.5 min per hour. The 7.5-min median concentrations were the basic units used to estimate SO_2 exposure.

The sulfur gas analyzer was calibrated against NBS-SRM 1627 permeation tubes using either a Model 303 gas mixing system (Analytical Instrument Development, Inc.) or a Model 330 Dynacalibrator (Metronics). Span checks with permeation tubes were conducted every 2 weeks. Multipoint recalibrations were conducted if the span check detected an error of 5% or greater. The need for recalibration was uncommon.

Sulfur dioxide concentrations were monitored at various plot locations (Figure 3.2) throughout the growing season. While not all locations on all plots were monitored, location c was continually monitored on all plots, providing a basis for interplot comparisons. Use of multiple locations within some plots allowed intraplot comparisons. Air samplers were usually placed 30–35 cm above the ground (approximate canopy height). In 1978, air samplers were placed at several heights to determine vertical gradients in SO_2 concentration.

The 7.5-min medians from each sampling location were used directly to estimate weekly, monthly, and seasonal geometric means (GM), standard geometric deviations (SGD), and arithmetic means (AM). By assuming that relative concentration among plots would remain the same when absolute values changed because of fluctuations in wind speed, continuous readings from each location could be estimated by normalizing and then aggregating the 7.5-min median concentrations (Preston and Lee, 1982). This approach was used to estimate the sequence of 1- and 3-hr averages.

Several potential sources of error in SO_2 monitoring data were recognized (Preston and Lee, 1982; Preston et al., 1980). The importance of these on the estimated concentrations can be gauged by comparing the maximum and minimum values that errors could produce. To facilitate the comparison, each data point was entered into two calibration equations. In the MAX run, the calibration equation was adjusted to yield a maximum concentration likely to result from including quantifiable sources of uncertainty. In the MIN run, the calibration equation was adjusted to yield the corresponding minimum estimate. Summary statistics were computed for both the MAX run and the MIN run. We assume that the range of values presented brackets the true values.

The time integrated horizontal distribution of SO_2 was determined by placing Huey sulfation plates at various locations at canopy height (35 cm) over the entire study site (Preston and Gullett, 1979). The plates, which absorbed ambient sulfur dioxide and retained it as lead sulfate, were placed in the field for 1 month, then collected and analyzed for sulfate content.

An extensive study indexed the general relative concentrations of SO_2 at various points in horizontal space on the plots. Plates were widely spaced over treatment

Table 3.1. Seasonal Average SO_2 Concentrations Measured on the Plots at Location c from 1975 to 1979 (MIN Run–MAX Run $\mu g \cdot m^{-3}$)[1]

		Site I				Site II			
Year	Measure	Control	Low	Medium	High	Control	Low	Medium	High
1975	GM[2]	25[7]	50[7]	90[7]	155[7]				
	SGD[3]	1.5	2.4	2.7	2.7				
	AM[4]	25	75	145	255				
	1-hr[5]	255	740	1245	2055				
	3-hr[6]	255	460	1085	1595				
1976	GM	5–20	40–50	85–90	150–155	5–30	60–65	95–100	150–160
	SGD	3.0, 1.5	3.6, 2.5	3.1, 2.7	2.9, 2.5	3.4, 1.2	2.3, 1.8	2.3, 1.9	3.0, 2.3
	AM	10–25	90	170	265	10–30	85	125	245
	1-hr	300	1410–1455	2935–3740	6100–8155	140	1570–1710	2540–2890	4595–5195
	3-hr	175–180	855–925	1895–2285	3950–4990	90	925–995	1550–1665	2700–3005
1977	GM	5–35	75–80	120–130	190–220	10–35	65–70	110–115	175–190
	SGD	4.5, 1.7	2.2, 2.1	2.7	3.1, 2.5	4.2, 1.2	2.1, 1.9	2.2, 2.1	2.7, 2.4
	AM	25–50	105–110	220–235	325–355	25–35	95–100	155–175	275–300
	1-hr	600–645	1180–1295	2265–2355	3165–3720	760–880	1525–1915	2840–3695	4410–5845
	3-hr	460–485	880–945	1480–1640	2220–2680	275–325	880–1040	1895–2285	3465–4295

1978	GM	25	65	120	200–205	20–30	60	105	210–215
	SGD	1.7, 2.1	2.7	3.1	3.2, 3.4	1.3, 2.3	2.3	2.6	2.8, 2.9
	AM	30	125	270	445	25–30	90	185	410
	1-hr	600	2540	3950–3975	10,650–10,740	140	695	1315	2890
	3-hr	255	1615	2520	6790–6815	90	510	1085	2055
1979	GM	20–30	55–60	115	170	20–30	55–60	115–120	220–225
	SGD	0.8, 1.6	2.4, 2.1	2.5	2.4	2.9, 1.3	2.6, 2.1	2.7, 2.5	3.0, 2.8
	AM	35–40	85	195	1030	30–35	85–90	195–200	420
	1-hr	440	1050–1095	2020–2025	3190–3200	450	920	1400	3530
	3-hr	305	635–640	1215–1220	2030–2060	240	595	1095	2625

[1] Single values are presented when MIN-Run and MAX-Run values rounded to the same value.
[2] Geometric mean.
[3] Standard geometric deviation.
[4] Arithmetic mean.
[5] Highest 1-hr average concentration observed during the growing season.
[6] Highest 3-hr average concentration observed during the growing season.
[7] MAX run only for 1975.

and interval plots. One plate was placed in the center of each quadrant of each plot and a fifth plate placed next to the real-time sulfur analyzer sample position c on experimental plots (Figure 3.2) and in the geometric center of the interval plots. The study was repeated on both Sites I and II for four test periods during 1977.

In an intensive study, we wished to determine the dispersion pattern of the SO_2 immediately upon leaving the delivery system. Sulfation plates were placed in a tight pattern around five gas delivery orifices on the low plot. The study was repeated for both Sites I and II for three test periods in 1977.

3.2.3 Temporal Patterns in SO_2 Concentrations

From the performance of a prototype of the delivery system (Lee et al., 1975), we expected concentration frequency distributions to be lognormal and that the 3-hr peaks on experimental plots would bracket the federal 3-hr secondary standard, 1300 $\mu g \cdot m^{-3}$. We hoped not to exceed any Montana or federal standards on the low treatment plots. For skewed distributions, such as the lognormal, the median is commonly used as the measure of central tendency rather than the AM because the latter is extremely sensitive to values in the tail of the distribution. For aerometric data, the AM usually falls near the thirtieth percentile rather than the fiftieth (Larsen, 1971).

For lognormal distributions, the median and geometric mean (GM) are equal, and the standard geometric deviation (SGD) characterizes temporal variability in concentrations around the geometric mean. Thus, taken together, the GM and SGD characterize dose.

For individual plots receiving direct SO_2 the highest seasonal GM's were approximately 50% greater than the lowest (Table 3.1). Peaks (3-hr and 1-hr) were more variable than the GM's. Fluctuations of this magnitude are typical of lognormal distributions. Geometric mean SO_2 concentrations and their standard geometric deviations on the treatment plots were comparable to those monitored in area sources (Table 3.2).

Sulfur dioxide concentrations were generally higher at night than during the day (Figure 3.3). All concentration frequency distributions were approximately lognormal. Cumulative frequency distributions for different treatment plots were generally distinct, whereas distributions for various points monitored within a treatment plot approximated each other. Separation of cumulative frequency distributions of different treatments plots was greatest near median concentrations and least at the extremely high and low concentrations (Lee et al., 1979).

Sulfur dioxide concentrations varied considerably during the course of each growing season (Figure 3.4). The variability increased with exposure concentration. The high plots were particularly variable (Lee et al., 1979; Preston et al., 1980).

Sulfur dioxide concentrations on the plots were higher and more variable during the nighttime hours than during daylight hours (Lee et al., 1979; Preston et al.,

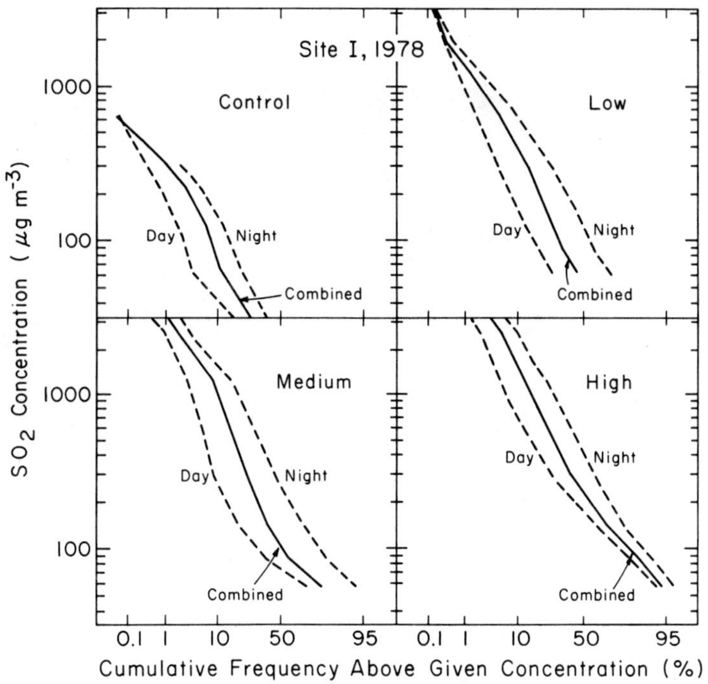

Figure 3.3 Day–night comparison of frequency distributions (7.5 min medians) of SO_2 concentrations ($\mu g \cdot m^{-3}$) on Site I, 1978 (MAX Run).

Table 3.2. Geometric Mean SO_2 Concentrations ($\mu g \cdot m^{-3}$) and their Standard Geometric Deviations for Several Urban Areas and the Site I and II SO_2 treatments[1]

Site	GM	SGD	1-hr Peak	24-hr Peak
Chicago	280	2.2	4530	2115
Site II–high	185–195	2.6–3.0	4410–5845	1155–1410
Site I–high	170–180	2.7–2.9	10,650–10,740	1384–1710
Philadelphia	150	2.4	2760	1235
Site II—medium	105–110	2.3–2.5	2840–3695	645–740
Site I—medium	105–110	2.8	3950–3975	740–830
Washington	105	2.2	1660	670
St. Louis	75	2.8	2575	695
Site II—low	60–65	2.0–2.4	1525–1915	345–370
Site I—low	55–60	2.4–2.7	2540	440–485
Cincinnati	45	2.9	1530	480
Los Angeles	35	2.3	775	270
Denver	35	2.1	965	160
Site II—control	10–30	1.3–3.3	760–780	70–80
Site I—control	15–25	1.5–2.8	600–645	115
San Francisco	15	2.9	695	215

[1]Data for 1975 (Site I) or 1976 (Site II) through 1979 are based on 7.5-min medians. Urban areas based on 5 min averages, 1962–1967 (HEW, 1970)

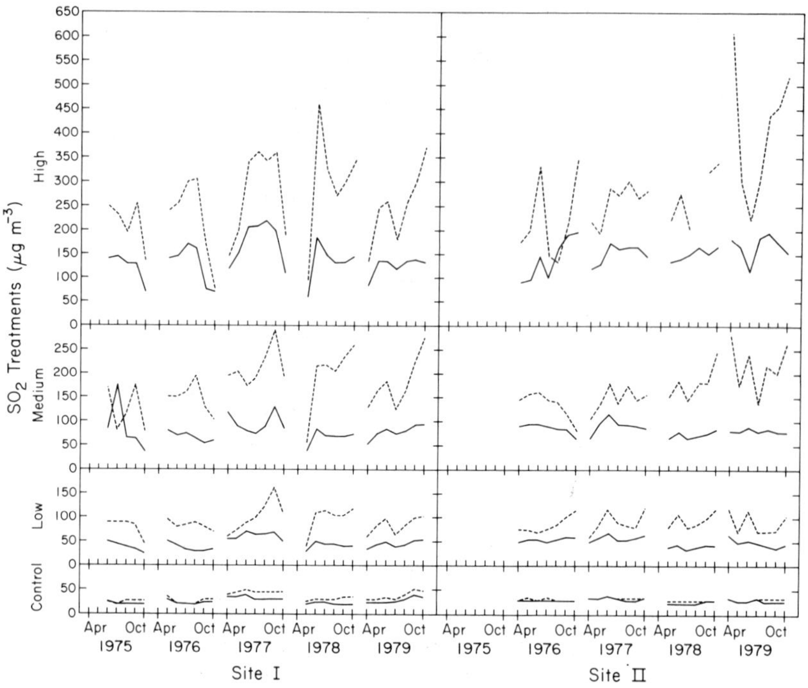

Figure 3.4. Daytime and nighttime monthly geometric mean SO_2 concentrations ($\mu g \cdot m^{-3}$) on site. —— Daytime, ---- Nighttime (MAX Run).

1980). This was caused by generally low wind speeds at night compared to relatively higher wind speeds during the daytime. Diel fluctuations in SO_2 concentrations generally followed one of four patterns (Figure 3.5):

1. A single concentration peak between 1800 and 2400 hr (Figure 3.5a);
2. A single peak concentration between 2400 and 600 hr (Figure 3.5b);
3. Two peak concentrations, one between 1800 and 2400 hr and one between 2400 and 600 hr (Figure 3.5c);
4. Low variable concentrations with no dominant peaks (less than 4% of days) (Figure 3.5d).

Occurrence of the first three patterns was strongly correlated with patterns of wind speeds (Lee et al., 1979). The fourth pattern usually occurred during days having high average wind speed and little variability.

3.2.4 Horizontal Spatial Distribution of SO_2 at the Canopy Surface

In the extensive study statistically significant differences in sulfation rates among locations occurred only on the high plots (Preston and Lee, 1982; Preston

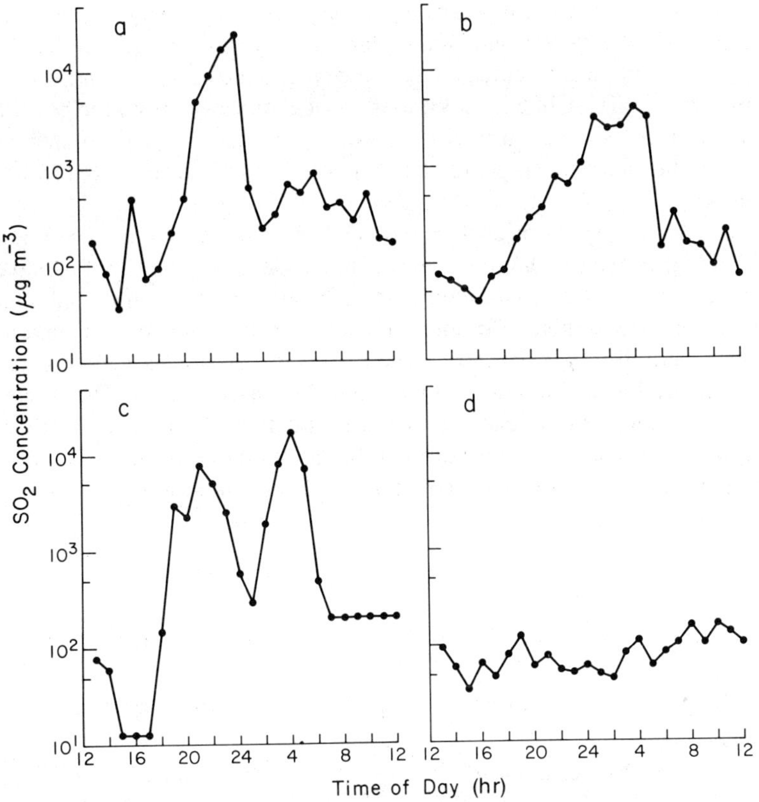

Figure 3.5. Four typical diel patterns (a–d, see text for discussion) in SO_2 concentrations ($\mu g \cdot m^{-3}$) on experimental plots.

and Gullett, 1979). Although other potential causes, such as system design, cannot be eliminated, the observed sulfation patterns can be explained by wind speed and direction (Preston and Gullett, 1979). The seasonal average concentrations on the control plots were about one fourth of those on adjacent low plots (Site I = 25.9%, Site II = 27.2%).

The horizontal dispersion of SO_2 appeared quite uniform on both sites beyond 1 m from the SO_2 release points (Preston and Lee, 1982).

3.2.5 Space-Time Variation of SO_2 Concentrations within the Vegetation Canopy

Canopy air flows can largely determine the rate of SO_2 transport to biota from the SO_2 delivery system. Horizontal winds flowing over the canopy surface are slowed by frictional drag on the vegetation. Immediately above the canopy, the mean horizontal wind velocity decreases logarithmically with decreasing height.

Within the canopy, a canopy–eddy layer exists in which wind speed decreases exponentially. In the lowest part of the plant–air layer, a logarithmic wind profile resumes with wind speed decreasing to zero at ground level (Inoue, 1963). The degree of vertical mixing of SO_2 at various vertical distances from the SO_2 delivery pipes on the plots should vary directly with wind speeds at those heights and therefore should decrease as wind speeds decrease with decreasing height above ground level.

Condition of the vegetation surface may also influence the vertical SO_2 gradient. Because of higher relative humidity and dew formation at night, SO_2 uptake rates greater than those during daytime may occur by adsorption even though nighttime stomatal resistance is high. Greater uptake rates at night should enhance the vertical concentration gradient. In general, sulfur dioxide concentrations were highest near the top of the plant canopy and decreased with height (Figure 3.6) (Wong et al., 1980). Significant vertical stratification of sulfur dioxide occurred primarily at night when concentrations of the pollutant reached maximum levels. During daytime, concentrations were lower and vertical stratification was slight.

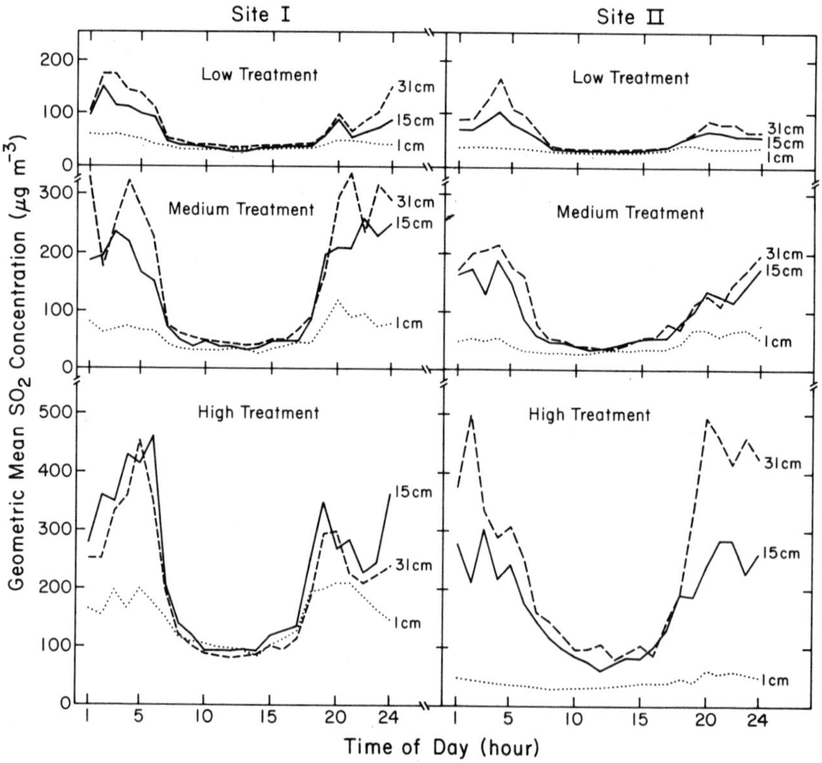

Figure 3.6. Examples of hourly SO_2 concentrations within the vegetation canopy at Sites I and II, 1978.

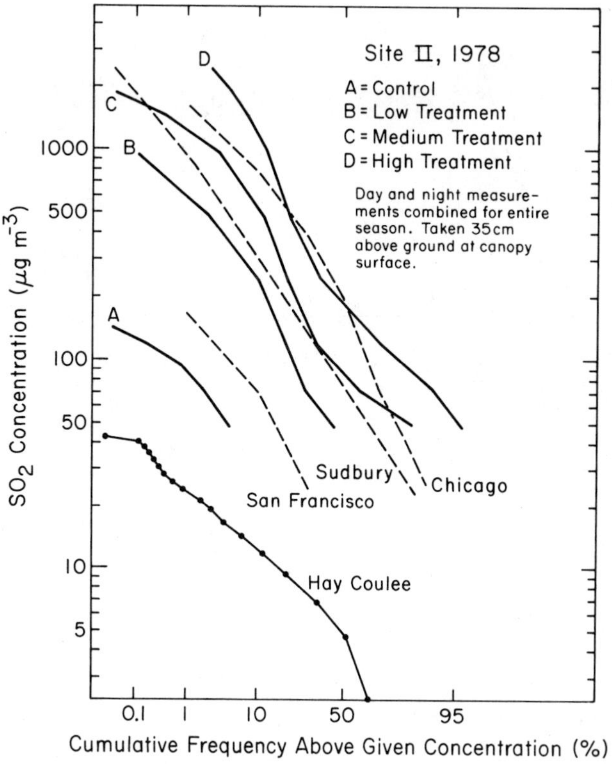

Figure 3.7. Frequency distributions of SO_2 concentrations monitored on Site II, 1978 (MAX Run) compared with those monitored in Chicago. San Francisco (1962–1967) (HEW, 1970), and Hay Coulee (7.12 km southeast of Colstrip power plants, 1977–1979), and Sudbury, 1964–1968. Hay Coulee data are from Ludwick et al. (1980). Sudbury data are from Dreisinger and McGovern (1970).

3.3 Comparison of Treatment SO_2 Concentrations with Actual Pollution Sources

3.3.1 Concentration Frequency Distributions

Concentration frequency distributions for Site II, 1978 are qualitatively similar (approximately lognormal with similar SGDs) to those with two major area sources (Chicago and San Francisco), a point source (two 330-megawatt (MW) coal-fired generating plants at Colstrip, MT) and the smelters at Sudbury, Ontario (Figure 3.7).

3.3.2 Diel Pattern of Concentrations

Area sources tend to have a bimodal pattern of daytime SO_2 concentrations with peaks at midmorning and late afternoon followed by low, constant nighttime

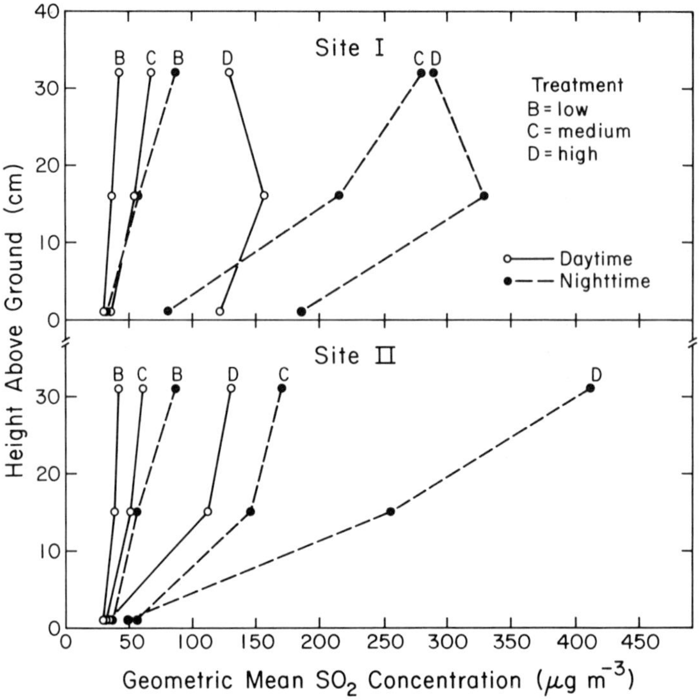

Figure 3.8. Vertical profiles of SO_2 concentrations (GM, $\mu g \cdot m^{-3}$) for daytime (6 a.m.–6 p.m.) and nighttime in 1978.

levels (Holzworth, 1973; Munn and Katz, 1959). These patterns are primarily determined by variations in mixing, wind speed and source strength. Elevated concentrations in the vicinity of a point source are likely to be more frequent during the daytime because conditions often favor mixing, while at night stable conditions tend to prevent mixing of the plume to the ground (Smith et al., 1979). This pattern was observed at Colstrip, Montana (Ludwick et al., 1980).

Though the exposue regime appears to simulate the shape of the concentration frequency distribution curve at Colstrip (Figure 3.7) reasonably well, it does not replicate the diel cycle of this point source. The treatment exposure regime is expected to simulate an area source with greater fidelity. Sulfur dioxide concentrations in an area-wide plume could be expected to have the same relationship with wind speed as those observed on our sites.

3.4 Biological Significance of SO_2 Exposure Patterns

Most organisms have diel activity cycles and consequently are likely to be affected more at certain times of day than at others. The biologically effective SO_2 exposure for green plants must be radically different from that for nocturnal rodents. In developing relationships between exposure concentrations and effects (dose–response), differences in biologically effective exposures among various organisms must be taken into consideration. In general, day-active organisms received substantially lower exposure at canopy height than did night-active organisms (Figure 3.8). The biologically effective exposures of day-active organisms may be substantially less than those implied by the overall seasonal summary data (Table 3.1).

Since SO_2 exposure concentrations are also vertically stratified within the vegetation canopy, quantification of the dose delivered to system components becomes complex. In a grassland canopy, components are also vertically stratified spatially and/or temporally. Vegetation would receive a variety of dosages throughout its height. Macroarthropod exposure would vary temporally depending on temporal patterns in the use of vertical space. Ground-dwelling organisms would generally receive the lowest overall direct dose.

References

Dreisinger, R. B., and McGovern, P. C. 1970. Monitoring atmospheric sulfur dioxide and correlating its effects on crops in the Sudbury area. In *Impact of Air Pollution on Vegetation Conference*, S. N. Linzon, ed. Ontario Department of Energy and Resource Management, Toronto, 1970.

Durran, D. R., M. J. Meldgin, Mei-Kao Liu, T. Thoem, and D. Hendersen. 1979. A study of long range air pollution problems related to coal development in the Northern Great Plains. *Atmos. Env.* 13:1021–1037.

Gifford, F. A., Jr. 1974. The form of the frequency distribution of air pollution concentration. In *Proceeding Symposium Statistical Aspects of Air Quality Data*. Research Triangle Park, NC: EPA-650 14-74-038. U.S. Environmental Protection Agency.

Heagle, A. S., D. E. Body, and W. W. Heck. 1973. An open-top field chamber to assess the impact of air pollution on plants. *Environ. Qual.* 2:365–368.

Health, Education, and Welfare, Dept. of. 1970. *Air Quality Criteria for Sulfur Oxides*. Washington, DC: U.S. GPO.

Holzworth, G. C. 1973. Variations of meterology, pollutant emissions, and air quality. In *Second Joint Conference of Sensing Environmental Pollution*. pp. 247–255. Washington, D.C.: American Chemical Society.

Inoue, E. 1963. On the turbulent structure of airflow within crop canopies. *J. Meteorol. Soc. Jap.* 41(6):317–325.

Knox, J. B., and R. I. Pollack. 1974. An investigation of the frequency distribution of surface air pollutant concentrations. In *Proceedings Symposium Statistical Aspects of Air Quality Data*. Research Trinagle Park, NC: EPA-65014-74-038. U.S. Environmental Protection Agency.

Larsen, R. I. 1971. A mathematical model for relating air quality measurements to air quality standards. Research Triangle Park, NC: EPA Report AP-89.

Lee, J. J., R. A. Lewis, and D. E. Body. 1975. The field experimental component: evaluation of the zonal air pollution system. In *The Bioenvironmental Impact of a Coal-Fired Power Plant*, R. A. Lewis, N. R. Glass, and A. S. Lefohn, eds. pp. 188–202. Second Interim Report Colstrip, Montana. EPA-600/3-76-013. Corvallis, Oregon: U.S. Environmental Protection Agency.

Lee, J. J., E. M. Preston, and D. B. Weber. 1979. Temporal variation in SO_2 concentration on ZAPS. In *The Bioenvironmental Impact of a Coal-Fired Power Plant*, E. M. Preston and T. L. Gullett eds. pp. 284–305. Fourth Interim Report, Colstrip, Montana. EPA-600/3-79-044. Corvallis, Oregon: U.S. Environmental Protection Agency.

Ludwick, J. D., D. B. Weber, K. B. Olsen, and S. R. Garcia. 1980. Air quality measurements in the coal-fired power plant environment of Colstrip, Montana. *Atmos. Environ*. 14:523–32.

Northern Great Plains Resources Program. 1975. *Effects of Coal Development in the Northern Great Plains*. Staff Report.

Munn, R. E., and M. Katz. 1959. Daily and seasonal pollution cycles in the Detroit-Windsor Area. *Int. J. Air Poll.* 2:51.

Pollack, R. I. 1975. Studies of pollutant concentration frequency distributions. EPA-650/4-75-004. Research Triangle Park, N.C.: U.S. Environmental Protection Agency.

Preston, E. M., and T. Gullett. 1979. Spatial variation of sulfur dioxide concentration on ZAPS during the 1977 field season. In *The Bioenvironmental Impact of a Coal-Fired Power Plant*, E. M. Preston and T. L. Gullett, eds. pp. 306–331. Fourth Interim Report, Colstrip, Montana. EPA-600/3-79-044. Corvallis, Oregon: U.S. Environmental Protection Agency.

Preston, E. M., T. L. Gullett, and D. B. Weber. 1980. Temporal variation in SO_2 concentrations on ZAPS during the 1978 field season. In *The Bioenvironmental Impact of a Coal-Fired Power Plant*, E. M. Preston and D. W. O'Guinn, eds. pp. 96–107. Fifth Interim Report, Colstrip, Montana. EPA-600/3-80-052. Corvallis, Oregon: U.S. Environmental Protection Agency.

Preston, E. M., and J. J. Lee. 1982. Design and performance of a field exposure system for evaluation of the ecological effects of SO_2 on native grassland. *Environ. Monitoring Assessment* 1:213–228.

Ripley, E. A., and R. E. Redmann. 1976. Grassland. In *Vegetation and the Atmosphere, Vol. 2. Case Studies*, J. L. Monteith, ed. pp. 349–398. London: Academic Press.

Singh, J. S., M. J. Trlica, P. G. Risser, R. E. Redmann, and J. K. Marshall. 1980. Autotrophic subsystem. In *Grasslands, Systems Analysis, and Man*. pp. 59–200. Cambridge: Cambridge University Press.

White, I. L., M. A. Chartock, R. L. Leonard, S. C. Ballard, M. W. Gilliland, E. J. Malecki, E. B. Rappaport, F. J. Calzonetti, M. S. Eckert, T. A. Hall, G. D. Miller, and M. D. Devine. 1979. *Energy From the West*. EPA-600/7-79-082. Washington, D.C.: U.S. Environmental Protection Agency.

Wong, L. T. K., Preston, E. M., and T. L. Gullett. 1980. Vertical SO_2 concentration profile on ZAPS during the 1978 field season. In *The Bioenvironmental Impact of a Coal-Fired Power Plant*, E. M. Preston and D. W. O'Guinn, eds. pp. 108–119. Fifth Interim Report, Colstrip, Montana. EPA-600/3-80-052. Corvallis, Oregon: U.S. Environmental Protection Agency.

4. Sulfur Deposition, Cycling, and Accumulation

D. G. Milchunas and W. K. Lauenroth

4.1 Introduction

The input of sulfur into terrestrial ecological systems occurs through atmospheric SO_2 washout, particulate rainout, gaseous dry deposition, absorption by plants, and the weathering of underlying parent material. Losses occur via leaching, runoff, and volatilization. In the semiarid grasslands of the Great Plains of North America, losses of sulfur via leaching and runoff can be assumed to be of little consequence. Exceptions may occur in a small area of the Texas Panhandle and Southeastern New Mexico on typic aridisol and torriorthent soils that have a horizon of gypsum and calcium sulfate. Volatilization of sulfur in unfertilized and sulfate-treated soils has been shown to be less than 0.05% (Banwart and Bremner, 1976). Therefore, we expect inputs of sulfur via our SO_2 treatments to be retained within the system. Exposure to SO_2 can be expected to alter the usual cycling patterns because of the potential for high-sulfur concentrations to affect processes that regulate flows and because of the different mode and location of entry of the sulfur into the system.

Sulfur accumulation in a system exposed to SO_2 can be assessed by quantifying the partitioning of sulfur among system components. The transfers, flow rates, and temporal dynamics involved in the component partitioning processes will be addressed in the second section of this chapter. In the first section we will examine the deposition and pools of sulfur in the vegetation, litter, and soil. The vegetation will be further subdivided into an aboveground component consisting of functional groups of plants and a belowground component consisting of crowns, roots, and rhizomes.

4.2 Standing Stocks of Sulfur

4.2.1 Pool Sizes

We computed sulfur pools based on peak plant biomass and sulfur concentration data. Since we did not find significant effects of SO_2 on plant biomass (see Section 5.3), sulfur pool data were calculated from mean maximum biomass across SO_2 treatments within each site. In this way, plot-to-plot variation in biomass was eliminated and the sulfur pools more accurately describe sulfur loading of the system in response to SO_2 exposure concentrations. The data will be presented (1) for the control plots on Sites I and II; these are then compared with nitrogen pools calculated from data of Bokhari and Singh (1975) for a similar mixed prairie at Cottonwood, South Dakota; and (2) for the SO_2 treatments, to evaluate sulfur loading as a result of exposure to SO_2.

4.2.1.1 Control Conditions

Cool-season grasses dominated the aboveground sulfur compartment with 0.12 $g\ S \cdot m^{-2}$, or 75% of the aboveground sulfur pool (Figure 4.1). Cool-season forbs comprised 19% of the aboveground sulfur pool. The majority of the sulfur in the autotrophic subsystem was located in roots. The 0- to 10-cm soil layer contained approximately 80% of the total root biomass, and this portion of the roots comprised 83% of the belowground plant sulfur pool and 65% of the combined above- and belowground plant sulfur pool.

The quantity of sulfur in aboveground plants was small compared with that of other components of the system (Figure 4.1). The amount of sulfur tied up in litter was 62% greater than the amount in aboveground plants at peak standing crop. The soil sulfur pool to a depth of 15 cm contained 88.5 $g\ S \cdot m^{-2}$ compared with only 0.26 $g\ S \cdot m^{-2}$ in litter and 0.75 $g\ S \cdot m^{-2}$ in plants (aboveground and belowground to a depth of 10 cm). Assuming litter occupies a 0- to 1-cm layer above the soil surface, litter would contain only 4% as much sulfur as the 0- to 1-cm soil layer.

Sulfur pools are compared with nitrogen pools for mixed prairies in Figure 4.2. Belowground plant components comprised 76% of the total autotrophic nitrogen pool, which is lower than the 79% for sulfur, considering that the nitrogen values are based on a 20-cm depth and sulfur to only a 10-cm depth. The amount of nitrogen tied up in litter was 95% greater than the amount in aboveground plants and is greater than the 62% computed for sulfur. A difference is also apparent in sulfur-to-nitrogen distributions within autotrophic belowground components. There was 16% of the belowground sulfur residing in the crowns compared with a value of 24% for nitrogen, even though root nitrogen was computed to 20 cm and root sulfur to only 10 cm.

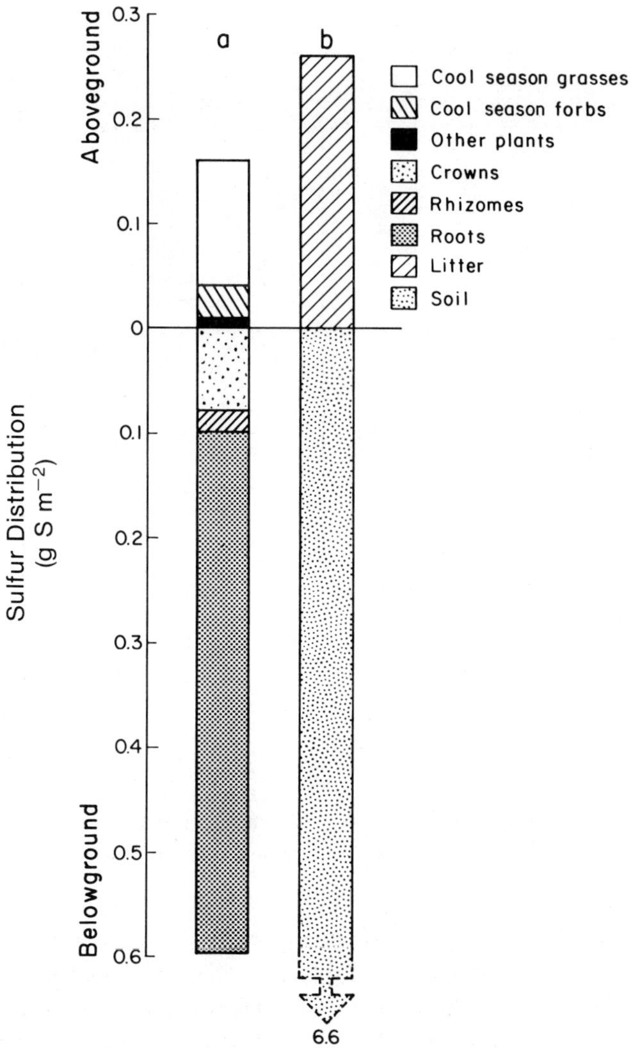

Figure 4.1. Sulfur pools (g S · m^{-2}) for a mixed prairie at peak biomass (not exposed to SO_2). (a) Plants aboveground by functional group and belowground by organ, and (b) litter and soil.

4.2.1.2 SO_2 Exposure

With the introduction of SO_2 as a source of sulfur to the system, aboveground sulfur pools increased on Site I by 57, 130, and 250%, and on Site II by 52, 140, and 330% for the low-, medium-, and high-SO_2 treatments (Figure 4.3). The proportion of the aboveground sulfur pool represented by functional plant groups was not affected by SO_2 treatment nor did it affect belowground plant sulfur pools.

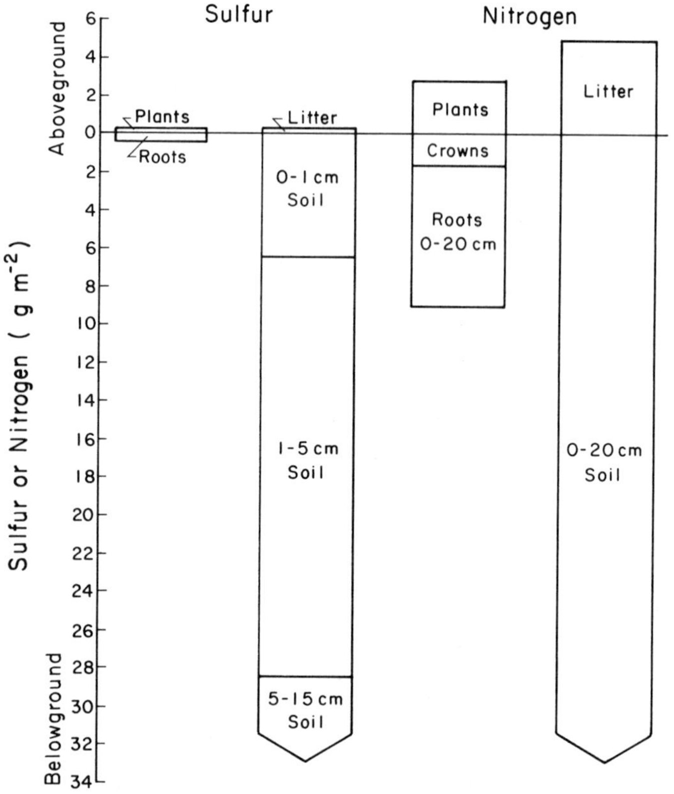

Figure 4.2. Sulfur (S) and nitrogen (N) pools (g · m^{-2}) for mixed-prairie sites (not exposed to SO_2). Nitrogen pools were compiled from data of Bokhari and Singh (1975).

Litter sulfur pools increased 20, 40, and 60% with exposure to the low-, medium-, and high-SO_2 treatments (Table 4.1). This was small compared with the increases in aboveground plant sulfur pools. There are two reasons for this. First, the decomposition of litter deposited during one growing season proceeds for several years. Therefore, the SO_2–sulfur absorbed by live vegetation that subsequently became litter entered a litter pool, of which a portion remained from the years of plant production prior to SO_2 exposure. It would take a period equal to

Table 4.1. Sulfur Pools (g S · m^{-2}) in Litter and Soil to a Depth of 1 cm for Site I

Fraction	Control	Low	Medium	High
Litter	0.20	0.24	0.28	0.33
Soil	6.11	7.12	6.43	7.56

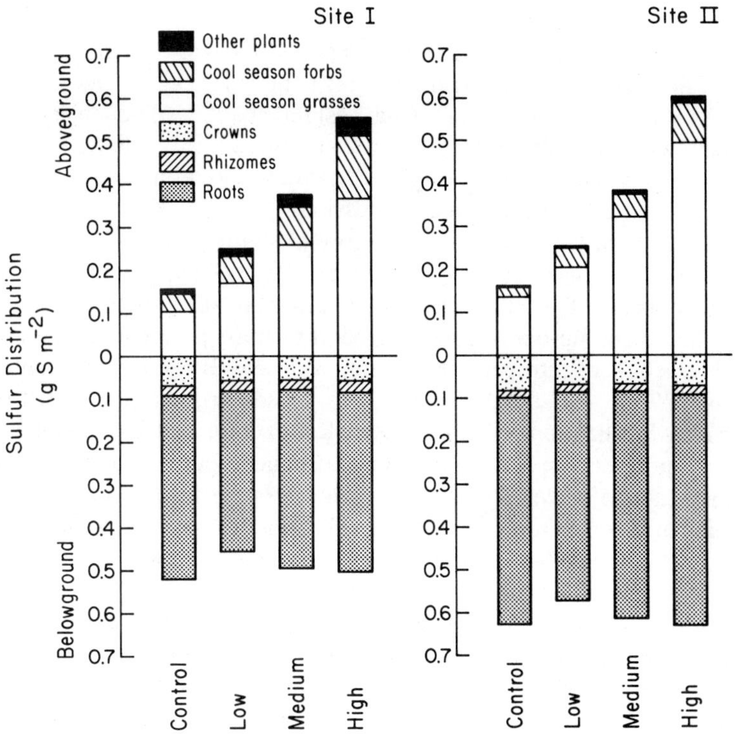

Figure 4.3. Sulfur pools (g S · m $^{-2}$) by aboveground plant functional groups and belowground plant organs for a mixed prairie at peak biomass exposed to three controlled levels of SO_2.

the turnover time of litter before deposition into an undiluted pool would be attained. Second, a large portion of the sulfur accumulated by live plants was released to the soil by the following year (see Section 4.3.5). Cell solubles, compared to the cellulose, hemicellulose, and lignin of structural tissue, have high-sulfur and nitrogen-to-carbon ratios and are also the component of plants that are most accessible to microbial utilization and subject to leaching.

The large quantities of sulfur incorporated into the soil via the death, decomposition, mineralization, and leaching of plant biomass were small relative to the size of the soil sulfur pools. Very large soil sulfur pools masked any increments in soil total sulfur due to the SO_2 treatments; however, pools of soil extractable sulfate (0–4 cm) increased from 0.4 g $SO_4^{=}$–sulfur · m^{-2} on the control to 0.5, 0.7, and 2.3 g $SO_4^{=}$–sulfur · m^{-2} on the low-, medium-, and high-SO_2 treatments. The small increments in sulfate pools on the low- and medium-SO_2 treatments probably represented a lag time for incorporation of $SO_4^{=}$ into organic or otherwise bound forms, whereas the large increase on the high-SO_2 treatment represented a buildup after the saturation of immobilization potentials. This topic is pursued in Section 4.3.5.

The interaction of SO_2 with soil sulfate and nitrate and the subsequent impact on aboveground *A. smithii* sulfur and nitrogen pools were assessed by fertilizing plots on each of the SO_2 treatments with the equivalent of 150 kg N $\cdot$ ha^{-1} (ammonium nitrate) and/or 15 kg S $\cdot$ ha^{-1} (magnesium sulfate). The objective of this experiment was to evaluate the interactions among SO_2–sulfur, sulfate–sulfur, and nitrogen on sulfur and nitrogen dynamics and biomass production (see Sections 5.5 and 5.6 for additional discussion).

Nitrogen fertilization increased aboveground sulfur pools by 0.01, 0.13, 0.13, and 0.10 g S $\cdot$ m^{-2} on the control, low-, medium-, and high-SO_2 treatments. These increases reflected the effect of the interaction of growth rate with SO_2 concentration on sulfur uptake. Aboveground biomass production was maximized by the interaction of nitrogen fertilization with low-SO_2 concentration (i.e., 96, 222, 154, 129 g $\cdot$ m^{-2} for the control, low-, medium-, and high-SO_2 treatments with nitrogen fertilization). With nitrogen fertilization and increasing SO_2 concentrations, more atmospheric sulfur was assimilated per unit leaf area, but total biomass production declined. Thus, the increase in the sulfur pool was similar on the low- and medium-SO_2 plus nitrogen fertilization treatment and then declined to control levels on the high-SO_2 with nitrogen fertilization treatment. No interaction of SO_2 with sulfate fertilization was observed.

The aboveground sulfur pools presented in Figure 4.3 underestimate total yearly sulfur loading to the system via live plant deposition by factors of the current year standing dead, by dead material already in the litter pool by late July, and by a small amount of growth after the peak biomass was measured. Adjusting the peak standing crop sulfur pools for current standing dead and losses to the litter gave values for total sulfur uptake by vegetation of 0.18, 0.27, 0.42, and 0.63 g S $\cdot$ m^{-2} $\cdot$ $year^{-1}$ on Site I, and 0.20, 0.30, 0.48, and 0.85 g S $\cdot$ m^{-2} $\cdot$ $year^{-1}$ on Site II for the control, low-, medium-, and high-SO_2 treatments (Table 4.2). The amount of

Table 4.2. Annual Sulfur Uptake (g S $\cdot$ m^{-2} $\cdot$ $year^{-1}$) by Various Components of a Mixed Prairie Exposed to Three Levels of SO_2, Site I 1976.[1]

	SO_2 Treatment			
	Control	Low	Medium	High
Peak live aboveground biomass	0.1598	0.2489	0.3747	0.5640
Recent dead[2]	0.0086	0.0115	0.0190	0.0261
Recent litter[2]	0.0116	0.0142	0.0261	0.0394
Total uptake by growing plants	0.1800	0.2746	0.4198	0.6295
Attributable to SO_2 treatments[3]	—	0.0946	0.2398	0.4495
Dry deposition to dead material	0.0035	0.0064	0.0074	0.0081
Dry deposition to bare soil	0.0042	0.0336	0.0701	0.1270
Total sulfur flux	0.1877	0.3146	0.4973	0.7646
Attributable to SO_2 treatments[3]	—	0.1269	0.3096	0.5769

[1]Values for dry deposition to dead plant material and to bare soil are derived from a simulation model based on Site I, 1976, (see Chapter 7); all other data are from field data.
[2]Assuming same S concentration when live as standing peak live.
[3]Amount above control levels.

sulfur uptake by vegetation that can be attributed to the SO_2 treatments is therefore 0.09, 0.24, and 0.45 g S · m^{-2} · $year^{-1}$ for the low-, medium-, and high-SO_2 treatments.

Sulfur dioxide can directly impact vegetation by accumulating in plant tissue, or indirectly by accumulating in the soil and thereby altering soil chemistry, decomposition, and other regulators of nutrient cycling processes. The degree to which direct or indirect effects of SO_2 operate depends in part on the productivity and turnover rate of the vegetation. Estimates of the deposition to dead plant material and to bare soil were obtained using simulation model (see Section 7.2.5) output based on 3-hr mean SO_2 concentrations and abiotic conditions for the 1976 growing season (Table 4.2). Deposition to dead plant material was of little consequence to the sulfur budget of the system and showed only small increments with increasing SO_2 concentration. Direct deposition of SO_2–sulfur to bare soil was only 0.004 g S · m^{-2} · $year^{-1}$ on the control but increased significantly across SO_2 treatments to 0.127 g S · m^{-2} · $year^{-1}$ on the high-SO_2 treatment. The increases in deposition to bare soil that were attributable to our low-, medium-, and high-SO_2 treatments were estimated to be 0.029, 0.066, and 0.123 g S · m^{-2} · $year^{-1}$. The percentage increases in sulfur deposited from SO_2, compared with the control levels, were 75, 77, and 78% for live vegetation uptake; 2, 2, and 1% for deposition onto dead plant material; and 23, 21, and 21% for deposition onto bare soil for the low-, medium-, and high-SO_2 treatments. Eriksson (1963) estimated global sulfur inputs from SO_2 to be 75×10^6 tons absorbed directly by plants and 25×10^6 tons absorbed by soil. Although these proportions closely agree with ours, differences could be expected in more humid environments because of the high solubility of SO_2 in water.

4.3 Dynamics of Sulfur

In this section we will examine the seasonal and yearly dynamics of sulfur in plants on the SO_2 treatments, the distribution within individual leaves, and the form of sulfur accumulated within the plant. The dynamics of sulfur in aboveground plant parts will be discussed in relation to nitrogen and the implications of the seasonal nitrogen-to-sulfur ratios to consumers. We will then follow the translocation of sulfur belowground and examine the form and concentration of sulfur in belowground organs. Finally, we will look at the influence of rainfall on plant sulfur concentrations, and the concentrations of sulfur in dead plant material, litter, and soil.

4.3.1 Sulfur and Nitrogen in Aboveground Plants

Sulfur dioxide treatments significantly altered the concentrations and dynamics of sulfur in *A. smithii* tillers (Figure 4.4). Whereas control tillers had lower sulfur concentrations at the end of the growing season compared with the peak concentrations in spring, tillers exposed to the low-SO_2 treatment had sulfur

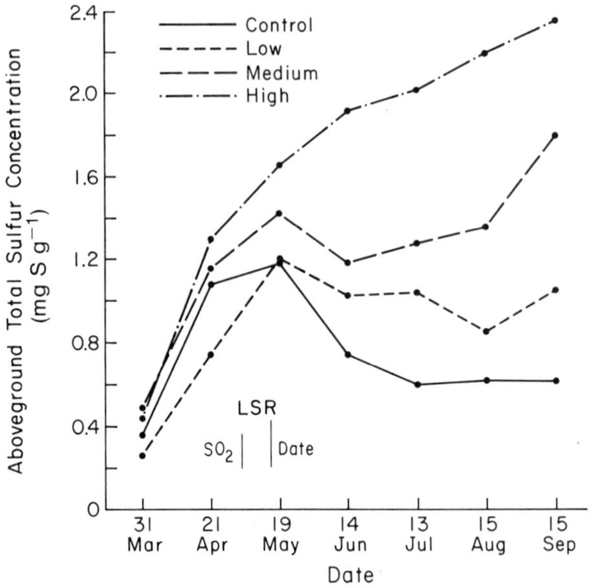

Figure 4.4. Sulfur concentrations (mg S · g^{-1}) in aboveground *A. smithii* tillers through a representative growing season, Site I, 1979.

concentrations that were similar to the spring levels through the growing season. Tillers growing on the medium-SO_2 treatment had higher sulfur concentrations by the end of the growing season, and tillers exposed to the high-SO_2 treatment showed an increase in sulfur concentration throughout the growing season.

The seasonal patterns of tiller sulfur concentration (Figure 4.4) were a function of the uptake of both soil and atmospheric sulfur. A large portion of the sulfur uptake early in the growing season was from the soil; i.e., 99, 84, and 73% for the low-, medium-, and high-SO_2 treatments assuming all sulfur in control tillers was obtained from the soil. After the spring peak, sulfur concentrations of control tillers declined and then remained relatively constant from June to August. We used June to September increments in plant sulfur concentration (ΔS) to describe sulfur uptake during the period of the growing season in which atmospheric uptake predominated as the mode of sulfur enrichment in the plants exposed to SO_2 (Milchunas et al., 1983).

The change in tiller sulfur concentration from spring to fall (ΔS) on Site I increased with increasing SO_2 concentration, but the rate of this increase decreased at high-SO_2 concentrations (Figure 4.5). The decrease in the rate of increase in tiller sulfur concentration with increasing SO_2 concentration indicates a decreased capacity for uptake at high-SO_2 concentration. This may reflect stimulated stomatal opening at low-SO_2 concentration and stomatal closure at high-SO_2 concentration (Majernik and Mansfield, 1970) and/or feedback inhibition of root sulfate uptake when shoot sulfur concentrations were high (Schiff and Hodson, 1973). The nonlinearity of uptake across SO_2 concentration cannot, however, be taken as a generality. Similar data for Site II were more closely linear.

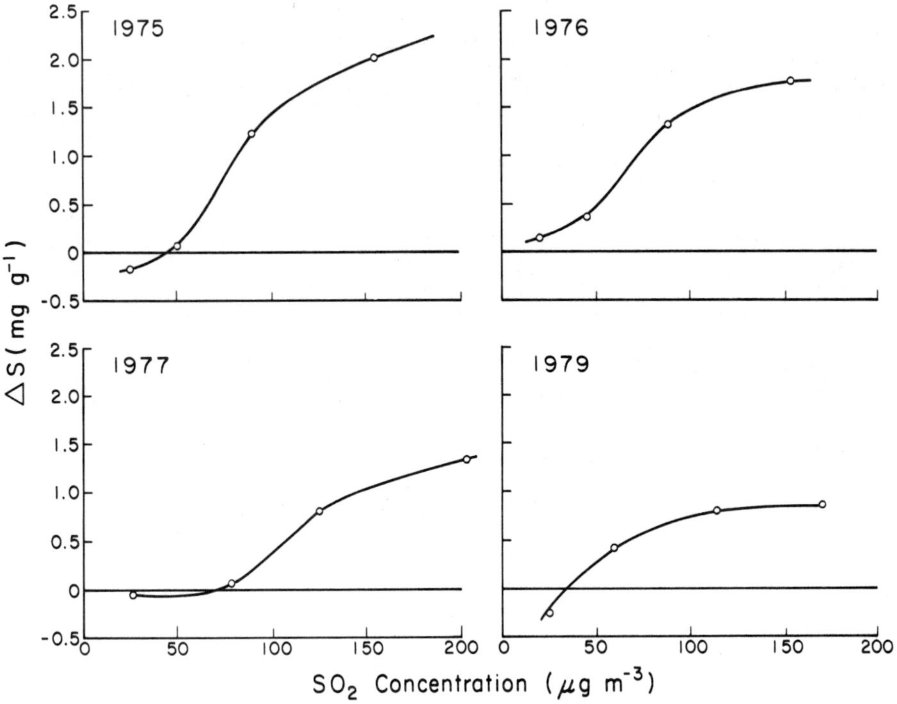

Figure 4.5. Increments (June to August) in the sulfur concentration (ΔS) of *A. smithii* tillers in relation to atmospheric SO_2 concentration ($\mu g\ S \cdot m^{-3}$) for 1975, 1976, 1977, and 1979.

Differences between Site I and Site II which may have influenced this relationship include greater primary productivity and the presence of green vegetation later in the growing season on Site II.

Sulfur dioxide exposure resulted in a reversal of the relationship between sulfur concentration in *A. smithii* tillers and yearly productivity that was observed on the control. Whereas sulfur concentrations of control tillers were low in years when productivity was high, sulfur concentrations of *A. smithii* tillers exposed to SO_2 were highest when productivity was highest (Table 4.3). Sulfur concentrations in August were 1.2, 0.9, and 1.2 mg $S \cdot g^{-1}$ for control tillers compared with 3.3, 4.1, and 3.2 mg $S \cdot g^{-1}$ for high-SO_2 treatment tillers in 1975, 1976, and 1977. Peak standing crops of *A. smithii* were 56, 69, and 39 $g \cdot m^{-2}$ in 1975, 1976, and 1977. Although greater biomass production dilutes nutrient concentrations in tillers not exposed to SO_2, the increased gas exchange associated with greater productivity allows for a more than proportionately greater uptake of SO_2. Low concentrations of SO_2 can stimulate plant growth by correcting nutrient sulfur deficiency (Faller, 1971; Cowling and Lockyer, 1978). This short-term positive effect can eventually be detrimental because of increased sulfur loading to the system.

Differences in sulfur accumulation appeared not only in sulfur concentrations due to different productivities between years, but also in total sulfur uptake due to

Table 4.3. Sulfur Concentration (mg S · g^{-1} plant) of *A. smithii* for 4 Months, Four Treatments, Two Sites, and 3 Years of SO_2 Treatment[1]

Site and SO_2 Treatment	Year	Month of Collection May	June	July	August	Treatment Mean[2]
Site I	1975					
Control		1.8[a]	1.0[a]	1.9[ab]	1.2[a]	1.4[a]
Low		1.3[a]	1.4[ab]	1.4[a]	1.3[a]	1.4[a]
Medium		1.4[a]	2.1[bc]	2.4[bc]	2.6[b]	2.2[b]
High		1.3[a]	2.3[bc]	2.8[c]	3.3[b]	2.4[b]
Site I	1976					
Control		0.7[a]	0.9[a]	0.9[a]	0.9[a]	0.9[a]
Low		1.1[ab]	1.4[a]	1.5[ab]	1.5[a]	1.5[b]
Medium		1.6[b]	2.2[b]	2.2[b]	2.9[b]	2.9[c]
High		2.4[c]	2.4[b]	3.2[c]	4.1[c]	3.2[d]
Site I	1977					
Control		1.3[a]	—	1.1[a]	1.2[a]	1.5[a]
Low		1.5[a]	—	1.3[a]	1.3[a]	1.5[a]
Medium		1.6[a]	—	2.2[b]	2.1[b]	2.2[b]
High		2.4[b]	—	2.7[b]	3.2[b]	2.9[c]
Site II	1976					
Control		1.3[a]	1.1[a]	1.0[a]	1.0[a]	1.1[a]
Low		1.4[ab]	1.2[a]	1.7[a]	1.5[a]	1.5[a]
Medium		2.1[b]	2.3[b]	2.7[b]	2.8[b]	2.5[b]
High		3.2[c]	3.5[c]	4.3[c]	4.8[c]	4.0[c]
Site II	1977					
Control		1.4[a]	—	0.9[a]	1.1[a]	1.2[a]
Low		1.7[a]	—	1.3[ab]	1.8[ab]	1.8[b]
Medium		1.9[ab]	—	1.8[b]	2.2[b]	1.9[b]
High		2.6[b]	—	3.8[c]	3.9[c]	3.3[c]

[1]Means not sharing common superscripts within a column for a site and year are significantly different ($P < 0.05$). Adapted from Milchunas et al. (1981a).
[2]Treatment means for a particular year by site may include data from April and September.

differences in productivity between Sites I and II within a given year. Peak biomass of *A. smithii* in 1976 was 60, 49, 48, and 55 g · m^{-2} on Site I and 95, 124, 85, and 105 g · m^{-2} on Site II on the control, low-, medium-, and high-SO_2 treatments. The total quantities of sulfur accumulated by *A. smithii* on Sites I and II (Figure 4.6) are correlated with differences in biomass between the two sites. The quantity of vegetation as well as the growth rate of the vegetation are important determinants of the sink strength of a system with respect to SO_2.

Defoliation treatments superimposed on the SO_2 treatments to simulate the additional impact of grazing also affected the concentrations and accumulation of sulfur by plants (Lauenroth et al., 1983) (see Section 5.6 for a description of the defoliation experiment). A single defoliation on 20 May resulted in similar rates of

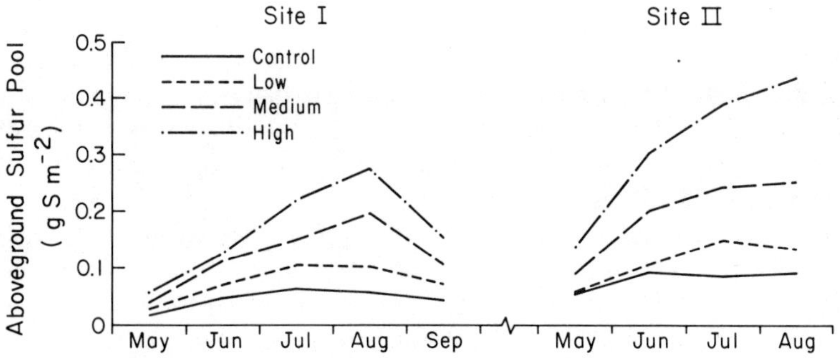

Figure 4.6. Total quantities of sulfur (g S · m^{-2}) in aboveground *A. smithii* tillers through the 1976 growing season.

increase in *A. smithii* sulfur concentration with increases in SO_2 concentration regardless of the degree of defoliation (Figure 4.7a). Compared with control tillers, one light defoliation treatment (50% of biomass) did not alter tissue sulfur concentration, whereas one heavy defoliation (100%) increased tiller sulfur concentration at all SO_2 concentrations. A second defoliation applied on 20 June altered the relationships among degree of defoliation with respect to the rate of increase in tiller sulfur content with increases in SO_2 concentration (Figure 4.7b).

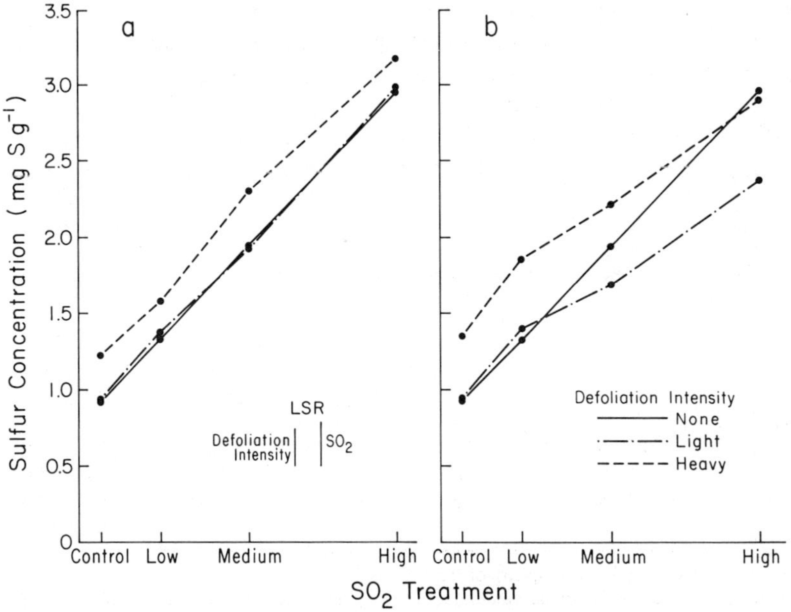

Figure 4.7. Sulfur concentrations (mg S · g^{-1}) of *A. smithii* tillers exposed to three concentrations of SO_2, two defoliation intensities, and (a) one or (b) two defoliations.

Sulfur content of tillers subjected to two heavy defoliations was significantly greater than that in control tillers at the control, low-, and medium-SO_2 treatments but not at the highest SO_2 treatment. In contrast, the sulfur content of tillers subjected to two light defoliations was not significantly different from the control tillers at the control and low SO_2 concentrations but was significantly less at the medium and high SO_2 concentrations.

The rate of sulfur uptake by *A. smithii* tillers was affected by the defoliation treatments (Table 4.4). Sulfur uptake nearly doubled with a single heavy defoliation at the control and low-SO_2 treatment concentrations and increased by 50% at the medium- and high-SO_2 concentrations. Reapplying the heavy defoliation decreased the rate of uptake to one third of that calculated for unclipped tillers. The higher sulfur concentrations and uptake rates, and similar biomass (see Figure 5.26), with a single heavy defoliation indicated that increased gas exchange occurred during regrowth which had the same effect on tiller sulfur content as increased productivity during wet years. Reapplying the heavy defoliation severely inhibited regrowth capacity and therefore decreased sulfur uptake rates. The rate of sulfur uptake of plants exposed to SO_2 can have an influence on the toxicity to the plant and to herbivores. Sulfur dioxide absorbed through stomates is converted to sulfite and then sulfate. Sulfite is more toxic to plants (Ziegler, 1975) and ruminant microorganisms (Trinkle et al., 1958) than sulfate. Greater rates of sulfur uptake may result in larger proportions of sulfite.

Nitrogen concentrations of *A. smithii* tillers (Table 4.5) displayed the same pattern as sulfur to different yearly productivities, i.e., lower concentrations when productivity was high. The seasonal dynamics of nitrogen are, however, quite different from those of sulfur. After initial high-nitrogen and -sulfur concentrations in spring, nitrogen continued to decline through the growing season while sulfur remained relatively constant. Nitrogen values varied by a factor of over 4, whereas sulfur values varied by only a factor of 2.

Exposure to any of the three SO_2 concentrations did not alter nitrogen concentrations of *A. smithii* tillers during any year or season. This is contrary to

Table 4.4. Sulfur Standing Crop (mg) and Uptake Rate ($mg \cdot day^{-1}$) for *A. smithii* Subjected to Defoliation and SO_2 Treatments

	Defoliation Treatment[1]				
		Light[2]		Heavy	
SO_2 Treatment	Control	Once	Twice	Once	Twice
Control	17 (0.125)	18	9	19 (0.218)	2 (0.036)
Low	21 (0.154)	26	13	25 (0.287)	3 (0.054)
Medium	24 (0.176)	23	11	25 (0.287)	4 (0.072)
High	44 (0.324)	48	22	36 (0.414)	6 (0.107)

[1]See Chapter 5, Section 5.6, for description of treatments.
[2]The rate of sulfur uptake could not be calculated for the light defoliation treatment because of the complication posed by the stubble.

Table 4.5. Nitrogen Concentration (mg N · g^{-1} plant) of *A. smithii* Tillers for Five Dates, Four Treatments, Two Sites, and 3 Years of SO_2 Exposure[1]

Site and Treatment	Year	Date: May	June	July	August	September	Treatment Mean[2]
Site I	1975						
Control		25.4	18.2	13.3	11.1	—	17.0
Low		25.1	17.2	13.7	10.8	—	16.5
Medium		26.9	16.8	13.6	11.2	—	15.7
High		26.6	19.4	11.8	10.0	—	17.0
Date mean		25.9[a]	17.9[b]	12.9[c]	10.8[d]	—	16.6
Site I	1976						
Control		22.3	15.4	12.4	9.7	7.6	15.0
Low		21.8	13.8	10.6	8.0	6.4	13.5
Medium		20.5	14.6	10.9	9.4	6.5	13.9
High		18.9	13.5	10.9	8.9	6.9	13.1
Date mean		20.9[a]	14.3[b]	11.2[c]	9.0[d]	6.8[d]	13.9
Site I	1977						
Control		22.1	19.8	13.8	12.7	9.4	17.1
Low		23.1	18.4	10.9	11.0	7.0	15.8
Medium		20.9	19.0	11.9	10.9	8.6	15.7
High		25.3	19.6	12.2	12.4	8.6	17.4
Date mean		22.9[a]	19.2[b]	12.2[c]	11.7[c]	8.4[d]	16.5
Site II	1976						
Control		22.9	11.0	13.8	10.4	9.8	14.5
Low		24.5	19.1	16.0	11.5	9.5	17.8
Medium		23.0	16.4	13.2	10.1	8.1	15.7
High		25.0	19.2	14.8	11.8	13.0	17.7
Date mean		23.8[a]	16.4[b]	14.4[b]	10.9[c]	10.0[c]	16.4
Site II	1977						
Control		25.9	22.1	15.8	17.3	—	20.3
Low		29.5	24.1	15.9	17.0	—	21.6
Medium		27.7	22.8	15.7	15.0	—	21.0
High		28.2	24.6	16.0	19.1	—	22.0
Date mean		27.8[a]	23.4[b]	15.9[c]	17.1[c]	—	21.2
All sites	All years						
Control		23.7	17.3	13.8	12.2	—	16.8
Low		24.8	18.5	13.2	11.7	—	17.0
Medium		23.5	17.9	12.8	11.3	—	16.3
High		24.8	19.3	13.1	12.4	—	17.4
Date mean		24.2	18.2	13.3	11.9	—	16.9

[1]Means not sharing a common superscript within a row are significantly different ($P < 0.05$). Adapted from Milchunas et al. (1981a).
[2]September data are not included in treatment mean.

reports of decreasing plant nitrogen concentrations with increasing SO_2 concentrations (Mishra, 1980; Garsed et al., 1981) or increasing nitrogen concentration with SO_2 exposure at low levels of N nutrition (Ayazloo et al., 1980). Cowling and Bristow (1979) did not detect differences in the total nitrogen concentration of plants exposed to SO_2 and grown in either sulfur-deficient or sulfur-supplemented soil, but did observe an increase in most free amino acids. Changes in free amino acid concentrations and enzymes involved in amino acid metabolism have been found to be a typical response to SO_2 (Jäger, 1975; Jäger and Klein, 1977; Wellburn et al., 1976; Godzik and Linskens, 1974; Malhotra and Sarkar, 1979). Increases in free amino acid content at high-SO_2 concentrations are brought about by protein hydrolysis (Godzik and Linskens, 1974; Malhotra and Sarkar, 1979). Thus, total nitrogen concentrations may not change while the form of nitrogen can be altered. Different responses in the total nitrogen concentration of plants exposed to SO_2 may be a function of the translocation and redistribution of nitrogenous compounds within the plant, which depends on the phenological stage and the nutritional status of the plant. There is a similarity in enzymatic and amino acid changes between the natural senescence process and SO_2 exposure (Milchunas et al., 1981b). Sulfur dioxide exposure induces earlier senescence of individual leaves, but tillers can partially compensate for this by increasing the production of new leaves (see Section 5.5). The increased turnover rates can result in lower nitrogen concentrations in individual older leaves with no change in the concentrations of the entire tiller. Our nitrogen concentration data represent values for entire tillers, whereas values of Garsed et al. (1981) are for old pine needles. Mishra (1980) reported values for the entire plant, but exposure concentrations were high and resulted in visible necrotic leaf lesions.

The different intraseasonal dynamics of sulfur compared with nitrogen, with and without SO_2, have important implications to the plant and to herbivores. Nitrogen and sulfur are utilized in plant and animal nutrition according to a stoichiometric relation. Dijkshoorn and Van Wijk (1967) reviewed published data on N:S ratios in the organic or protein fractions of various plant materials and proposed optimal N:S ratios of about 17.5:1 for legumes and 13.6:1 for grasses. Ruminants require a N:S ratio of 11–13:1 and rumen microorganisms of 12–14:1 (Leibholz and Naylor, 1971; Moir et al., 1967; Whanger et al., 1978). Narrower ratios, 10:1 or slightly less, in the diets of ruminants may lead to substantial improvements in the utilization of nitrogen (Moir et al., 1967), while ratios greater than 17:1 may be detrimental to animal health (Metson, 1973). High concentrations of sulfur have been shown to diminish animal intake rates (Bouchard and Conrad, 1974) and to be toxic to rumen microorganisms (Hubbert et al., 1958; Trinkle et al., 1958).

Nitrogen:sulfur ratios of *A. smithii* tillers that were not exposed to SO_2 were greater than 17:1 on four out of five May sample dates (Table 4.6). Sulfur dioxide exposure narrowed the N:S ratios and thus compensated for the inability of root uptake of sulfur to keep pace with that of nitrogen during rapid spring growth (Pitman, 1976; Dijkshoorn, 1959) as well as improve the quality of forage for herbivores. However, the continual decline in nitrogen through the growing season compared to relatively constant sulfur levels on the control and the increased sulfur

Table 4.6. Nitrogen to Sulfur Ratios for *A. smithii* for 4 Months, Four Treatments, Two Sites, and 3 Years of SO_2 Treatment[1]

Site and SO_2 Treatment	Year	Month of Collection				Treatment Mean[2]
		May	June	July	August	
Site I	1975					
Control		24	19	7	10	12
Low		20	12	9	8	12
Medium		19	8	6	4	7
High		20	8	4	3	7
Site I	1976					
Control		31	17	13	11	17
Low		19	10	7	5	9
Medium		13	7	5	3	6
High		8	6	3	2	4
Site I	1977					
Control		17	—	12	10	12
Low		15	—	8	8	10
Medium		13	—	6	5	7
High		11	—	4	4	6
Site II	1976					
Control		18	10	14	10	13
Low		17	15	10	8	12
Medium		11	7	5	4	6
High		8	6	3	2	4
Site II	1977					
Control		19	—	18	16	17
Low		11	—	12	10	12
Medium		14	—	9	7	11
High		11	—	4	5	7

[1]Adapted from Milchunas et al. (1981a).
[2]Treatment means for a particular year by site may include data from April and September.

concentrations on SO_2 treatments through the growing season resulted in near optimum N:S ratios on the control plots during July, August, and September, and very low N:S ratios (2–5:1) on SO_2 treatments for the same periods. Although we could not detect a significant effect on these very low N:S ratios on the *in vitro* digestibility of *A. smithii* (Milchunas et al., 1981a), this does not rule out the possibility of altered volatile fatty acid proportions or of effects independent of the digestive process (see Chapter 6, Ruminants). The general short-term intra-seasonal pattern that emerges for both plants and herbivores is one of positive effects of SO_2 exposure in the early growing season, and negative effects in the latter part of the growing season depending on the concentration of SO_2.

4.3.2 Differential Uptake by Various Species

Plants are known to differ in their sensitivity to SO_2 (Ziegler, 1975; Davis and Wilhour, 1976) and this may be partially related to sulfur uptake rates (Garsed and Read, 1977). Thus, the species composition of a plant community may influence sulfur accumulation in the system. Low concentrations of SO_2 decrease stomatal resistance (Unsworth et al., 1972; Biscoe et al., 1973; Black and Unsworth, 1980), whereas high concentrations of SO_2 may close stomates (Caput et al., 1978). The stomatal response is dependent on plant species and SO_2 concentrations, but is also influenced by humidity or vapor pressure deficit. Sulfur dioxide may stimulate stomates to open in moist air and to close in dry air (Majernik and Mansfield, 1972; Black and Unsworth, 1980). Black and Unsworth (1980) postulated that SO_2, once in the substomatal cavity, enters the tissue via epidermal cells adjacent to the guard cells and at low concentrations leads to a reduction in turgor in these cells and consequently to stomatal opening. In vapor pressure deficit-sensitive species, increased transpiration induced by SO_2 may lead to stomatal closure at large vapor pressure deficit levels.

Exposure of plants on the control and high-SO_2 treatments to $^{35}SO_2$ in July and September suggested a stimulation of stomatal opening on the high-SO_2 treatment (Coughenour et al., 1979). Certain species were more active in $^{35}SO_2$ uptake than others. Noteworthy were the exceptionally high mean relative SO_2 deposition velocities ($cm^3 \cdot g^{-1} \cdot hr^{-1}$) for *Taraxacum officinale* (64) and *Achillea millefolium* (31) and the exceptionally low values for *Antennaria rosea* (5), *Aristida longiseta* (3), and *Artemisia frigida* (4). Sulfur uptake rate for *A. smithii* was intermediate to these. The differential uptake rates of species within a given SO_2 treatment can be a function of stomatal resistance, growth rate, position of the plant in the canopy, and various morphological characteristics of the leaf. The vertical profile of SO_2 concentrations indicated relatively lower SO_2 concentrations near the surface of the ground compared with 30 to 60 cm above the ground (see Section 3.4). Decreasing wind speed at lower strata in the canopy increases boundary layer resistance which decreases SO_2 uptake. Further, the removal of SO_2 by foliage high in the canopy decreases the concentrations reaching lower strata. Thus, decumbent, slow-growing species such as *Antennaria rosea* would be expected to absorb less sulfur than *Achillea millefolium* or *Taraxacum officinale* which grow rapidly and attain large leaf areas high in the canopy.

Differences in the sulfur concentration between plant species (Figure 4.8) were similar to differences in uptake rates with the exception of *Artemisia frigida*. The high-sulfur concentrations with low atmospheric uptake rates of *A. frigida* must have been the result of greater uptake of sulfur from the soil. Relative to other species growing on the control treatments, *A. frigida* plants contained higher concentrations of sulfur. This could be related to higher lignin and volatile oil content of this aromatic half-shrub. Some plants contain appreciable amounts of sulfur in compounds other than sulfates, amino acids, thiamine, biotin, and coenzyme A. These include sulfides, disulfides, polysulfides, sulfoxides, and methyl sulfonium (Salisbury and Ross, 1969). The primary function of sulfonium compounds is in the transfer of their methyl groups to other compounds including

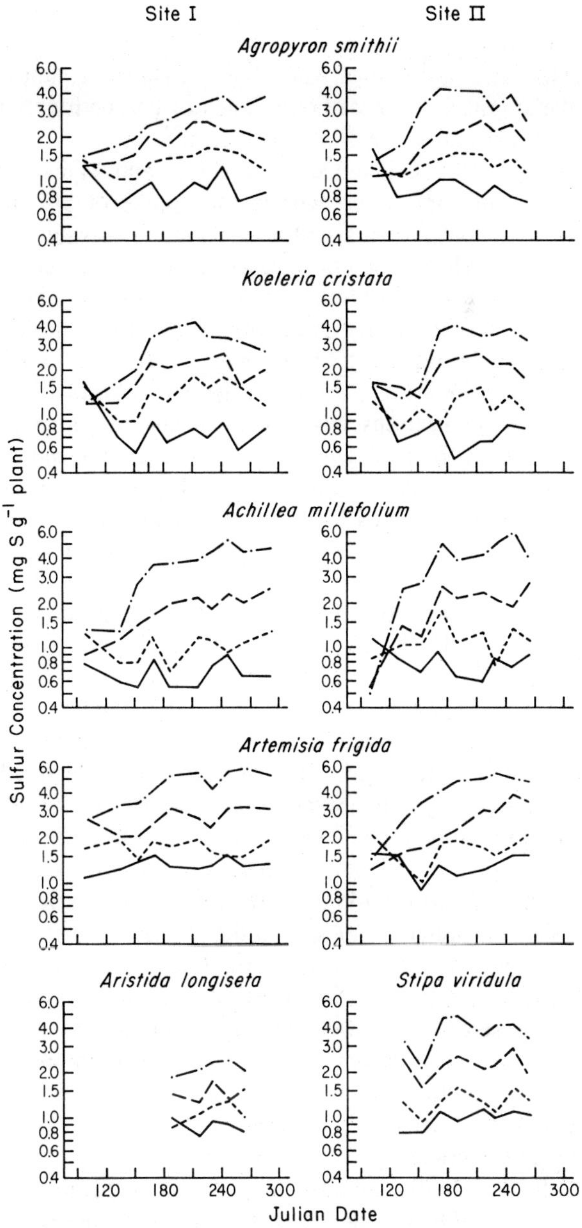

Figure 4.8. Sulfur concentrations (mg S · g^{-1}) in live tissue of five plant species, 1976. Control (——), low (----), medium (——), high (·—·). (Adapted from Rice et al., 1979.)

various alkaloids and the lignins. Alkaloids are allelopathic, and lignin encrusts cellulose and hemicellulose rendering it less available to soil and ruminant microbial breakdown. It is not known whether increasing the supply of sulfur to plants increases the synthesis of these compounds or competitively favors plants that produce high concentrations of allelochemicals.

Carlson and Bazzaz (1982), Klein et al. (1978), and Winner and Mooney (1980) suggested that, in part, differences in the uptake of SO_2 between plant species may be related to C_4 and C_3 photosynthetic pathways. C_4 plants are characterized by a high photosynthetic capacity, even when stomatal conductances are low, and low-water requirements, whereas the inverse applies to C_3 plants. Susceptibilities may also differ due to the competitive inhibition of ribulose 1,5-diphosphate carboxylase by sulfite (Ziegler, 1972). *A. smithii, Koeleria cristata*, and *Stipa viridula* are C_3 grasses, whereas *Aristida longiseta* is a C_4 grass. Similar sulfur concentrations were found in all the above grasses on the control plots (Figure 4.8). With SO_2 exposure, however, the much lower sulfur concentrations in *A. longiseta* lend support to the hypothesis that C_3 versus C_4 plants may be differentially selected for with respect to SO_2 exposure.

Another comparison between the sulfur concentration of different species involves the unique metabolic adaptation of certain species of *Astragalus* to accumulate very high concentrations of toxic selenium. In nonadapted plants, selenium substitutes for sulfur in the sulfur amino acids. Se–Se bonds are less stable than the usual S–S bonds, and this may result in a loss of enzyme activity and death of the plant (Harborne, 1977). *Astragalus* plants incorporate the selenium into Se-methylselenocysteine and selenohomocysteine which are nontoxic to the plant but toxic to ruminants after deamination and resynthesis of toxic selenomethionine and selenocysteine by microbial populations in the rumen. Since selenium and sulfur are readily interchangeable in the sulfur amino acids, and since some *Astragalus* species are selenium accumulators, we were interested in answering the following questions: (1) are *Astragalus* plants also sulfur accumulators? and (2) does atmospheric and soil sulfur inhibit or competitively reduce selenium uptake in a species that does not accumulate and process large quantities of selenium (Milchunas et al., 1983)? Fertilization of subplots on the SO_2 treatments with sulfur, selenium, and sulfur plus selenium indicated no inhibition of selenium uptake in *A. smithii* tillers exposed to SO_2, but competitive inhibition of selenium uptake on sulfur-fertilized plots (see Section 6.5.1). *Astragalus crassicarpus* plants accumulated very large amounts of sulfur. Sulfur concentrations of *A. crassicarpus* averaged 2 $mg \cdot g^{-1}$ on control plots and 7 $mg \cdot g^{-1}$ on the high-SO_2 treatment. The 228% increase in sulfur concentrations of *A. crassicarpus* plants exposed to SO_2 was, however, similar to the increases observed for other species. The high-sulfur concentrations of *A. crassicarpus* apparently result from an adaptation for the accumulation of large quantities of sulfur from the soil, and do not inhibit additional uptake of soil sulfur or selenium, or of atmospheric sulfur.

The uptake of SO_2, or tissue sulfur concentration, is an incomplete measure of differential species sensitivity to SO_2 because plants differ in the production of various sulfur compounds, secondary transport and storage mechanisms, internal morphology and chemistry, and metabolic pathways which sulfur may affect. After absorption of SO_2, sensitivity is partially a function of the ability to assimilate the pollutant derivatives and metabolize them to less harmful or benign forms. *A. smithii* accumulated substantial quantities of sulfur when exposed to SO_2 but cannot be considered sensitive to SO_2 (Lauenroth et al., 1979). High-sulfur concentrations in species of a community in the absence of SO_2 may indicate an ablity to utilize relatively large quantities of sulfur in the synthesis of compounds other than required nutrients. Thus, *Artemisia frigida* and *Astragalus crassicarpus* obtained high-sulfur concentrations on our control plots which may stem from the use of sulfur in production of volatile oils, alkaloids, or sulfur amino acid analogs.

Secondary translocation has been implicated as a pollutant tolerance mechanism based on observations of extremely high concentrations in the tips and margins compared to basal and medial leaf sections of some species (Compton and Remmert, 1960; Hill, 1969) but not in others (Guderian, 1977). A possible explanation may be that for a given leaf the boundary layer is not uniform over its entire surface but is thickest at the midvein and progressively thinner toward the margins (Salisbury and Ross, 1969). This would allow for greater transpiration from leaf margins and a tendency for solute movement toward this area, as well as influencing the location of SO_2 absorption. The possibility of high-suflur concentrations in the leaf margin being solely a function of deposition is not supported by data from our site which shows a similar phenomenon in *A. smithii* leaves that were and were not exposed to SO_2 (Table 4.7). Higher sulfur concentrations occurred in leaf tips compared to leaf bases in both young and old leaves, indicating independence from leaf age. A potential explanation that applies to *A. smithii* is that excess sulfate is sequestered in vacuoles. Since tips and margins represent the oldest tissue in grass leaves, the increase in sulfur concentration is a result of greater duration of exposure. An adaptive significance of this mechanism could be to isolate injury to the oldest portions of leaves while protecting other areas.

Translocation of sulfur absorbed as SO_2 to belowground organs and exudation from roots may be means of disposing of excess sulfur. Jensen and Kozlowski (1975) exposed seedlings of four forest tree species to $^{35}SO_2$ and found bigtooth aspen, the most sensitive species, translocated the lowest amount of ^{35}S, and sugar maple, the least sensitive species, translocated the most. There is no conclusive evidence that sulfur absorbed from the atmosphere and translocated to roots is readily leached or diffused into the soil. However, other investigators have presented data supporting this concept (de Cormis et al., 1969; Freid, 1948; Thomas and Hill, 1937).

Sulfur dioxide indirectly affects plants through the addition of sulfur to soils via direct absorption of the gas by soils (Smith et al., 1973; Lockyer et al., 1978; Payrissat and Beilke, 1975), deposition in rainfall (Baker et al., 1977; Nyborg et al., 1977), and the decomposition and mineralization of vegetation containing SO_2 absorbed sulfur. Acidification of soils is the main effect of these depositions. Soil acidity influences plants in a number of ways. Depending on species, they may suffer from increasing levels of soluble Al, Mn, and H, and deficiencies of Ca and

Table 4.7. Organic and Inorganic Sulfur Concentrations (mg S · g^{-1} leaf) of Young and Old *A. smithii* Leaf Sections on the Control and High-SO_2 Treatments

	Leaf Age							
	Young				Old			
	Base		Tip		Base		Tip	
	Organic	Inorganic	Organic	Inorganic	Organic	Inorganic	Organic	Inorganic
C[1]	0.32	0.52	0.55	0.57	0.45	0.53	0.59	0.61
H[2]	0.50	1.31	0.12	2.52	0.53	3.36	0.55	4.46

[1]C = Control.
[2]H = High SO_2.

Mg (Baker et al., 1977). The availability of N, P, K, and Mo is also lowered with decreasing soil pH (Buckman and Brady, 1969). Lowered soil pH decreases N fixation (Rice et al., 1977) and heterotrophic activity (Grant et al., 1979), thereby decreasing nutrient and carbon turnover in the system. Competitive relationships among plant species may be altered because of different nutrient requirements and nutrient uptake responses to changes in soil pH.

4.3.3 Form of Sulfur in the Aboveground Plant

Sulfur dioxide entering through stomates is assumed to be dissolved to form H_2SO_3, HSO_3^- and $SO_3^=$. Sulfite is then oxidized to sulfate (Garsed and Read, 1977). Sulfate is either metabolically converted to organic forms or stored. The partitioning of sulfur into sulfate sulfur and organic sulfur in plants exposed to SO_2 can thus indicate the extent to which incoming atmospheric sulfur is utilized in metabolic processes and also indicate when toxic levels reduce the synthesis of organic compounds or induce nutrient hydrolysis and plant senescence.

The sulfate–sulfur content of aboveground *A. smithii* tillers was affected by SO_2 treatment and changed with date of sampling (Figure 4.9). The sulfate–sulfur and organic sulfur responses were different from those of total sulfur in two important respects; early growing season increases were not as rapid for sulfate as for total sulfur, and total sulfur concentrations increased with increasing SO_2 concentration, while organic sulfur increased on the low- and medium-SO_2 treatments and then decreased on the high-SO_2 treatment. Organic sulfur averaged 89% of total sulfur in early spring (similar for all treatments), but by late in the growing season had declined to 55, 46, 36, and 5% for the control, low-, medium-, and high-SO_2 treatments. These responses can be attributed to differences in the conversion of sulfate to organic sulfur.

The different seasonal patterns between total, sulfate, and organic sulfur indicate a subsidy–stress gradient (Odum et al., 1979) along both seasonal and SO_2 concentration axes. The spring peak in the synthesis of organic sulfur compounds associated with rapid growth creates a demand for sulfur that SO_2 exposure apparently satisfies. As the season progresses, sulfate–sulfur concentrations increase proportional to SO_2 treatment. Toward the end of the season, the higher

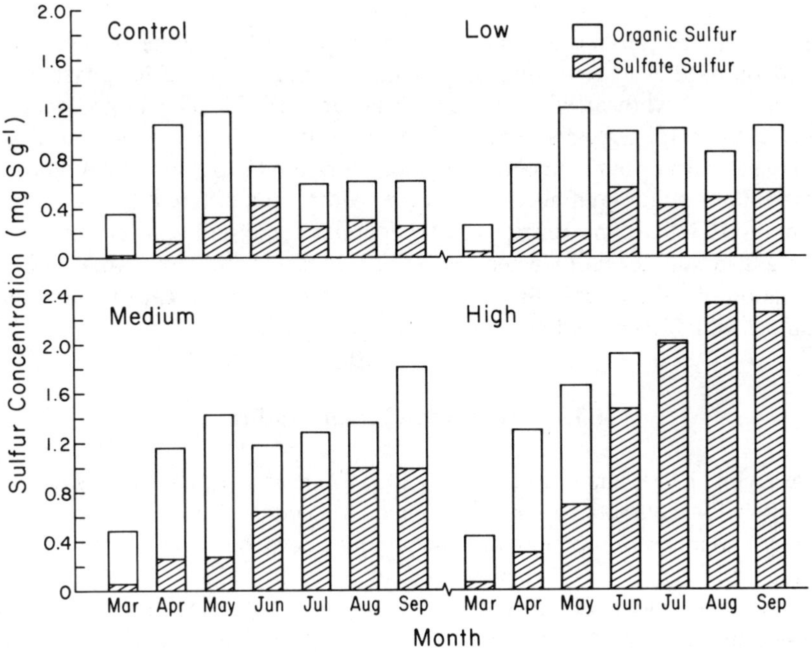

Figure 4.9. Sulfate–sulfur and organic sulfur concentrations (mg S · g^{-1}) of *A. smithii* tillers through the 1979 growing season, Site I. Sulfate–sulfur: $LSR_{SO_2} = 0.08$ mg · g^{-1}, $LSR_{Date} = 0.10$ mg · g^{-1}; organic sulfur: $LSR_{SO_2} = 0.13$ mg · g^{-1}. $LSR_{Date} = 0.16$ mg · g^{-1}.

organic sulfur concentrations on the low and medium-SO_2 treatments compared to control levels and the lower organic sulfur concentrations on the high-SO_2 treatment compared to control levels indicate a subsidy response from low- and medium SO_2 treatments but stress response on the high-SO_2 treatment.

The observed sulfur dynamics, when viewed with the concurrent responses of increased rates of leaf senescence, increased carbon translocation, reductions in leaf chlorophyll content (see Section 5.2.2) and decreased sulfur translocation (Section 4.3.4), pose interesting questions concerning senescence and the redistribution of nutrients. The senescence process in plants entails a hydrolysis of organic compounds and a general mobilization of chlorophyll, amino acids, RNA, lipids, carbohydrates, and nutrients (dela Fuente and Leopold, 1968; Scott and Leopold, 1966; Williams, 1955; Brady, 1973). There have been conflicting views on the mobility of sulfur in the plant which may stem from not differentiating between mobility of sulfate to shoots after root uptake and the redistribution mobility of sulfur out of leaves. Bouma (1975) presented an excellent review on the mobility of sulfur in relation to redistribution. Redistribution is quite extensive for nitrogen and phosphorus, but very low for sulfur. Further, it would be logical from an adaptive basis to assume preferential translocation of cytoplasmic, chloroplast, and nucleic material rather than material stored in the vacuole. Selective breakdown and translocation of cytoplasmic protein before chloroplast protein has been observed in plants with an adequate supply of sulfur to the roots (Hanson et al., 1941), but in sulfur deficient plants a depletion of storage sulfate occurred followed by a breakdown of cytoplasmic protein while chloroplast protein continued to be synthesized (Torii and Fujiward, 1967). The large decline in organic sulfur in July, August, and September on the high-SO_2 treatment may be related to senescence but cannot be altogether attributed to a redistribution process because sulfate–sulfur continued to increase. Although a different type of stressor was employed, Wood and Barrien (1939) found a decrease in protein–sulfur and an increase in sulfate–sulfur when leaves of ryegrass were kept in the dark. No translocation of sulfate took place to stems and roots, in contrast to amino acids and amides which increased greatly in roots and stems.

4.3.4 Belowground Plant Sulfur

Data of Kylin (1960) and Garsed and Read (1977) suggest that recently absorbed SO_2–sulfur is metabolized in preference to endogenous vacuolar sulfate. Transport and incorporation into organic products does, however, seem to occur in the same way regardless of whether the leaves are provided with SO_2, $SO_3^=$, or $SO_4^=$ via roots or shoots (Ziegler, 1975). Compounds are then used in metabolic processes, stored, or mobilized for export. If incoming $^{35}SO_2$–sulfur is not routed to storage vacuoles before it is translocated belowground, then ^{35}S label would not be diluted by the existing sulfur pool and the distribution of the label in belowground organs is indicative of the propensity or ability to translocate the SO_2–sulfur out of leaves.

Five days after exposing 0.5-m^2 plots to $^{35}SO_2$ for 5 hr (Coughenour et al., 1979), the percentage of ^{35}S in all harvested live material (above- and belowground) that was translocated belowground in July was similar on the control (9%) and the high-SO_2 treatment (10%). Sulfur dioxide treatment effects were, however, apparent when the $^{35}SO_2$ exposures were repeated in September. At this time, 15% of the ^{35}S was detected belowground on control plots and only 8% on the high-SO_2 treatment. Less sulfur was translocated belowground on the high-SO_2 treatment (Figure 4.10) even though aboveground sulfate concentrations were many times greater, and aboveground organic sulfur concentrations were lower, than control values. Comparative values for the translocation of sulfur to belowground organs are 5% in aspen to 24% in sugar maple seedlings after 8 days (Jensen and Kozlowski, 1975), and a range of 1.4 to 2.6% for tobacco (Faller, 1971).

The distribution of $^{35}SO_2$–sulfur among rooting depths and belowground organs was affected by date, but not SO_2 treatment. Roots to a depth of 5 cm were important sulfur sinks and their sink strength increased as the season progressed, receiving 66% in July and 80% in September of the ^{35}S translocated belowground. Less sulfur was translocated to roots at a depth of 5–10 and 10–20 cm in September than in July. Translocation to crowns and rhizomes was unaffected by date, with an average of 11 and 5% in each respective organ. The translocation of ^{35}S was in marked contrast to that of ^{14}C (see Section 5.4).

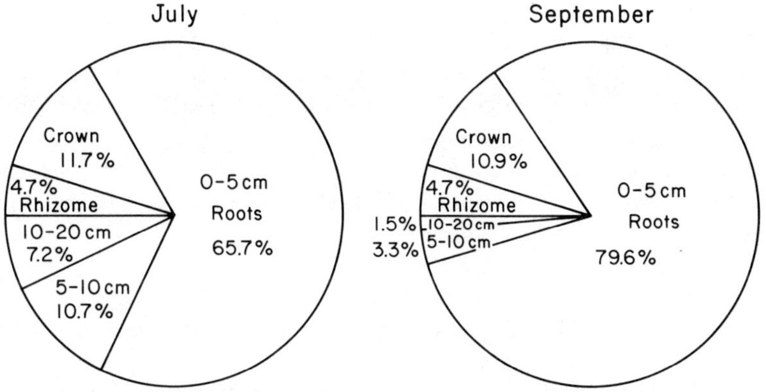

Figure 4.10. The distribution of $^{35}SO_2$-sulfur in belowground plant parts at two dates during 1976 at Site I. (Adapted from Coughenour et al., 1979.)

Contrary to results for total sulfur concentrations in aboveground plant parts, no significant SO_2 treatment or date effects were observed for total sulfur concentrations in roots or rhizomes of *A. smithii*. Sulfur concentrations of control tillers were slightly higher for crowns and roots compared with rhizomes or aboveground tillers (Table 4.8). Significant increases with increased SO_2 concentration were observed, however, for sulfate–sulfur concentrations of roots and rhizomes. Crowns were the only plant part that did not increase in either total sulfur or sulfate–sulfur with SO_2 treatment. The sulfate–sulfur concentrations of rhizomes decreased through the growing season, but no distinct trend was observed for the seasonal sulfate–sulfur concentrations of roots.

The lack of a significant SO_2 treatment effect on total sulfur concentrations of roots when shoot–sulfur concentrations were much higher was unexpected but is

Table 4.8. Ash-Free Total Sulfur and Sulfate–Sulfur Concentrations (mg S · g^{-1}) of *A. smithii* Exposed to Three Levels of SO_2 for 5 Years[1]

SO_2 Treatment	Control	Low	Medium	High
		Total Sulfur		
Aboveground tillers	0.69[a]	0.96[b]	1.30[c]	1.88[d]
Crowns	0.94[a]	0.93[a]	0.90[a]	0.93[a]
Roots	0.83[a]	0.73[a]	0.80[a]	0.81[a]
Rhizomes	0.66[a]	0.64[a]	0.61[a]	0.77[b]
		Sulfate–Sulfur		
Aboveground tillers	0.29[a]	0.41[a]	0.70[b]	1.62[c]
Crowns	0.20[a]	0.19[a]	0.21[a]	0.20[a]
Roots	0.15[a]	0.19[ab]	0.23[ab]	0.27[b]
Rhizomes	0.18[a]	0.20[ab]	0.22[ab]	0.26[b]

[1]Values represent means across May, June, July, August, and September, 1979, sampling dates. Means within a row that do not share a common superscript are significantly different ($P \leqq 0.05$).

consistent with the ^{35}S translocation results. Sulfate fertilization of subplots on the SO_2 treatments produced no significant effects on plant growth (see Section 5.5) or on sulfur uptake. No trends in sulfur uptake were observed within SO_2 treatments for either sulfate fertilization compared with controls or for nitrogen + sulfate fertilization compared with nitrogen fertilization alone. Endogenous reduced sulfur has a pronounced inhibitory effect on the rate of sulfate uptake by plant roots (Schiff and Hodson, 1973). Our data indicate that the feedback inhibition of root sulfur uptake in *A. smithii* is tightly controlled.

The increase in root sulfate concentrations with no increase in total sulfur concentrations can be explained only be considering plant sulfur metabolism and translocation. Photosynthesizing leaves function as the main centers for reduction of sulfate, whereas roots can reduce only a small portion of the sulfate they absorb (Rennenberg et al., 1979; Pate, 1965). On morphological, anatomical, and physiological grounds, a plant may be considered to be two systems, the root and the shoot, which perform contrasting and complementary functions. Brouwer (1963) termed the relationship between metabolites of the root and shoot a functional equilibrium. This functional equilibrium between roots and shoots in the case of sulfur consists of a circulation within the plant whereby roots supply shoots with sulfate and shoots supply roots with organic sulfur compounds. The control of this circulation appears to be mediated by the shoots, as evidenced by the control of sulfur uptake by roots when shoot concentrations are high. The decrease in sulfur translocation to belowground organs on the high-SO_2 treatment indicates curtailed circulation of sulfur compounds within the plant. The inability of roots to synthesize enough organic sulfur for its needs, and the decreased sulfur uptake and translocation from roots and organic sulfur translocation from leaves, resulted in a shift in the proportions of sulfur compounds, but not in the quantity of total sulfur, in roots. The lack of a feedback control on plant uptake of atmospheric sulfur results in high and often toxic concentrations of sulfur in the shoots, and also appears to disrupt the normal sulfur cycling relationship between roots and shoots.

4.3.5 Litter and Soil Sulfur

Sulfur dioxide–sulfur absorbed by plants eventually reaches the soil–sulfur pool via decomposition and mineralization of dead plant material, or from the leaching by rainfall of sulfur in live plants. Rainfall can leach considerable quantities of SO_2–sulfur that has been absorbed or adsorbed by leaves (Nyborg et al., 1977; Baker et al., 1977) as well as sulfur taken up by roots and translocated to leaves (Tukey, 1970). The structure of the mixed-prairie canopy at our site prevented the measurement of sulfur in throughfall versus openfall. However, the loss of sulfur from live and dead *A. smithii* tillers after rainfall events was determined to be insignificant in this semiarid habitat (Figure 4.11). Date, which represents duration of exposure, was the best single explanatory variable ($r^2 = 0.68$) for sulfur content of live tillers exposed to SO_2. This concurs with the results of

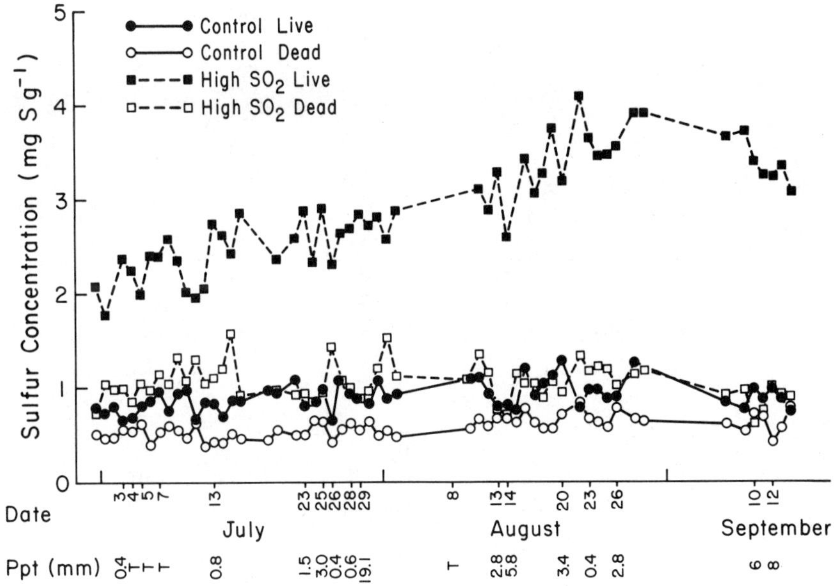

Figure 4.11. Sulfur concentrations (mg S · g^{-1}) in live and dead (previous year's) *A. smithii* tillers in relation to precipitation (mm, or T = trace) for Site I, 1979.

Lauenroth et al. (1979). Total rainfall for either 3 or 5 days preceding a sampling date increased prediction capabilities by only 2%, and the use of two additional variables (amount of rain in the last rain event and the number of rain days in the previous 5 days) increased the coefficient of determination by an additional 4% ($r^2 = 0.74$).

Decomposition and mineralization of dead plant tissue are therefore the principal means by which sulfur absorbed by mixed-prairie vegetation reaches the soil sulfur pool. The difference between the sulfur concentration of live tillers and standing dead tillers of the previous year (Figure 4.11) indicated a substantial overwinter loss of sulfur. The following year, losses of sulfur from dead tillers proceeded at a rate similar or slightly less than the loss of other constituents. The large initial overwinter loss of sulfur was apparent for a variety of plant species (Figure 4.12). A comparison of the sulfur concentrations of live tissue in Figure 4.8 (Section 4.3.2) with that for dead plant tissue of the current growing season in Figure 4.12, for the same species at a similar date, indicated that there was negligible translocation of sulfur from senescing leaves (redistribution). Sulfur absorbed by live vegetation during a given year of exposure in this grassland is, therefore, transferred to either litter or soil by the following year rather than retained in vegetation as residual sulfur. The sulfur content of live vegetation thus represents sulfur absorbed for a single year and not cumulative sulfur uptake with additional years of exposure.

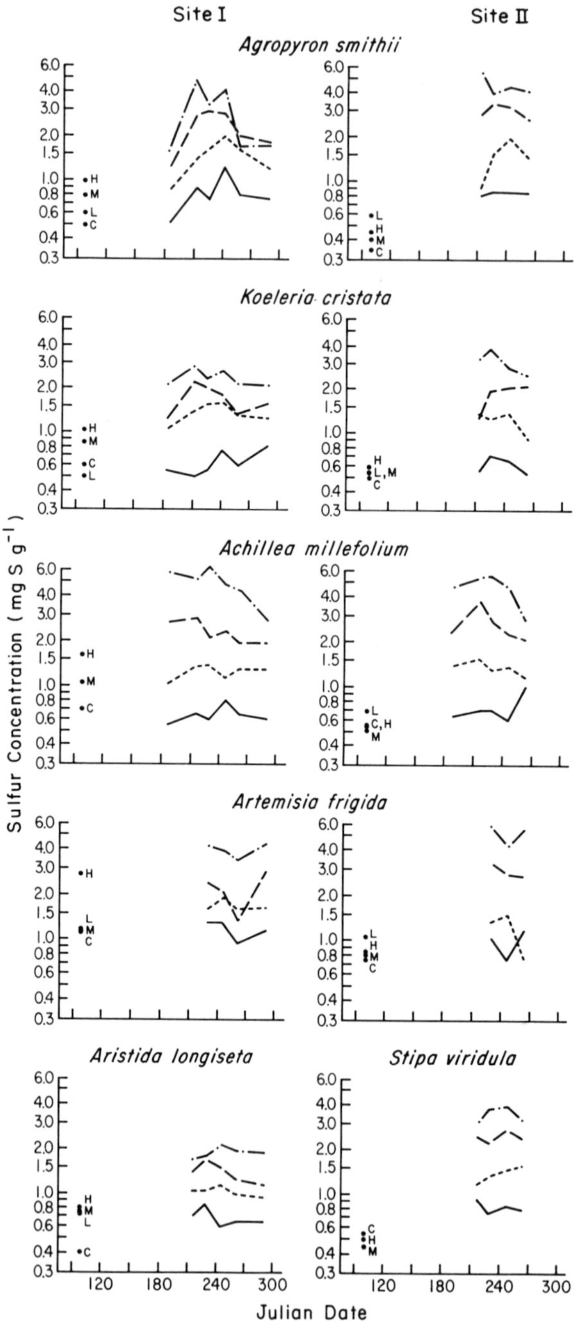

Figure 4.12. Sulfur concentrations (mg $S \cdot g^{-1}$) in dead aboveground plant parts for five species, 1976. Lines are current year's dead, dots are previous year's dead. Control (—), low (---), medium (——) high (.—.). (Adapted from Rice et al., 1979.)

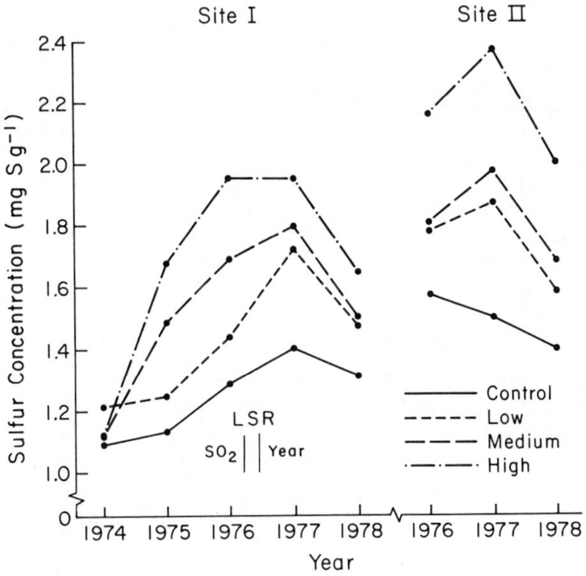

Figure 4.13. Sulfur concentrations (mg S $\cdot$ g^{-1}) in litter exposed to three concentrations of SO_2, 1974–1978. Exposure to SO_2 began in 1975 on Site I and in 1976 on Site II.

Total sulfur concentrations in litter were significantly affected by SO_2 treatment and changed with year of treatment but not over dates within a year. The sulfur concentrations of litter on Site I were significantly greater than control levels on the medium- and high-SO_2 treatments by the first year and on all treatments by the third year of exposure (Figure 4.13). Litter sulfur concentrations on Site II were significantly greater than control levels during the first year of treatment. The decline in litter sulfur concentrations observed in 1978 can be attributed to a very dry 1977 growing season and a very wet spring in 1978. We could not dctcct any cumulative trends in litter sulfur attributable to subsequent SO_2 exposure either across dates within a year or across years. Data from laboratory experiments indicated a rapid saturation rate for the dry deposition of SO_2 onto *A. smithii* leaf surfaces. It is probable that the rates of the dry deposition of SO_2 to litter parallel the rate of transfer of sulfur from the litter to the soil. Dry deposition would be expected to be of greater importance in humid environments because of the high solubility of SO_2 in water (Hocking and Hocking, 1977).

One important effect of elevated sulfur levels in litter is its potential influence on decomposition rates. Much of the aboveground net primary production passes through the decomposition process at ground level (Redmann and Abouguendia, 1978), and effects on litter decomposition rates may affect nutrient cycling processes. Two approaches were taken to study the rate of disappearance of *A. smithii* litter. The first examined 2 years of field data, and the second simulated field conditions in the laboratory. Since *A. smithii* was such a substantial component of aboveground biomass, we assumed that it would be representative of litter disappearance.

Table 4.9. Rate of Loss (mg · g^{-1} · day^{-1}) of *A. smithii* Litter for Five Time Intervals in 1978

Interval	Set I[1]	Set II[1]	Set III[1]
15 April–18 May	2.5	1.8	1.4
18 May–16 June	3.1	3.3	2.1
16 June–15 July	3.2	2.8	2.1
15 July–15 August	2.1	2.3	1.2
15 August–16 September	1.7	2.1	1.5

[1] See text for explanation.

In the 1978 field study, three sets of litterbags (Witkamp and Van der Drift, 1961) were placed on the experimental plots on 15 April. Set I contained material grown in the previous year on the control plot and was placed on the control plot. Set II was also placed on the control plot, but contained material from the high-treatment plot of the preceding year. Set III material had been grown on the high-SO_2 treatment plot and was placed on the high-treatment plot. Table 4.9 contains the summary of the results of this experiment as reported by Dodd and Lauenroth (1981) for the intervals between collection of the field-placed litterbags. Rate of loss was significantly lower for Set III, the samples grown and placed on the high-treatment plot, than either of the control sets, except for the first sampling for Set II, which had been grown on the high treatment and placed on the control.

In 1979, samples were collected from the control plot in March and placed on each of the three treatment plots on 15 April. A collection of bags on 14 June gave 93, 94, 96, and 97% of the initial material remaining on the control, low-, medium-, and high-SO_2 treatment plots, respectively, and the expected inverse relationship between SO_2 exposure and decomposition rate. The average rate of loss over this period was only 1.2 mg · g^{-1} · day^{-1} as opposed to the 1978 rate of 2.8 mg · g^{-1} · day^{-1}. The lower overall rate of loss in 1978 was attributed to drier soil surface conditions during June of that year. By 15 July, 88, 90, 89, and 89% of the initial material remained, respectively, as above, with notably diminished variation across treatments.

The laboratory decomposition study utilized samples collected in May 1979 from *A. smithii* litter of the previous growing season on a site that had not been exposed to SO_2. Samples were ground and placed on sand and exposed to a constant 228 $\mu g \cdot m^{-3}$ of SO_2 or a control airstream. Inhibition of decomposition rates in the material exposed to SO_2 was less during the first third than in the last two thirds of the 1-month exposure period. A 9% difference versus a 17% difference in decomposition rate was reported for these intervals (Leetham et al., 1983). This strongly suggests a cumulative effect of exposure to SO_2, since the SO_2 concentration was held constant. This cumulative effect is suggested to be due to the accumulation of SO_2 derivatives such as sulfate or sulfite and/or a reduction in pH of the litter-microbe system.

With the decomposition and mineralization of plant litter, and via direct dry deposition, sulfur from SO_2 is incorporated into the soil. However, no treatment-related differences were found in total sulfur concentrations of the soil on Site I

Table 4.10. Sulfur Concentration ($mg\ S \cdot g^{-1}$) of Soil to a Depth of 1 cm after 5 Years of Exposure to SO_2[1]

	Soil Sulfur Concentration ($mg\ S \cdot g^{-1}$)		
	1979	1980	
SO_2 Concentration	Site I	Site I	Site II
Control	0.44	0.53	0.44
Low	0.37	0.41	0.72
Medium	0.37	0.42	0.60
High	0.48	0.56	0.64

[1]Values represent a mean of four and six determinations for each SO_2 treatment for 1979 and 1980, respectively; and each determination is a pool of four soil samples.

(Table 4.10). Soil sampling was increased in 1980 on Site I and extended to Site II in an attempt to establish a relationship between soil sulfur and SO_2 exposure. We observed the same pattern between the soil sulfur concentration and the SO_2 treatments on Site I as in the previous year. Data from Site II displayed a different pattern. The already very large pools of soil sulfur compared with the inputs of sulfur via vegetation or litter (see Figures 4.1 and 4.3, Sections 4.2.1.1 and 4.2.1.2) masked any treatment effects on total sulfur.

The deposition of sulfur to soils as measured by soluble sulfates and as indicated by soil pH did, however, show SO_2 treatment effects. Sulfate concentrations of 0.007, 0.010, 0.013, and 0.045 $mg\ SO_4–S \cdot g^{-1}$ were found in the surface 4-cm soil samples taken from Site I in 1979 (Leininger and Taylor, 1981). The small increase over control sulfate concentrations on the low- and medium-SO_2 treatments and the very large increase on the high-SO_2 treatment suggest that a conversion and/or immobilization process was in operation which became saturated at the high-SO_2-treatment levels. Three possibilities exist. First, sulfate can be adsorbed by mineral colloids (Barrow, 1975). The sulfate thus adsorbed is only slowly available to plants and is not easily extractable (Nyborg, 1978). Second, sulfate can be incorporated into organic sulfur compounds by microbial activity in the soil. In the majority of soils, organic sulfur comprises over 95% of the total sulfur in the soil (Fitzgerald, 1978). We found 98% of the total sulfur in the soil was in an organic form on our sites. This suggests a degree of competition between plants and microorganisms for available sulfur. Further, there is evidence indicating that a relatively low C:S ratio is required by microorganisms for the maximum decomposition of plant cell walls because of increased sulfur metabolism involved in the production of extracellular enzymes necessary for the breakdown of cellulose (Freney and Swaby, 1975). Third, sulfur may form nonextractable apparently organic sulfur by a reaction with soil organic matter (Nyborg, 1978). Once these sulfate conversion and immobilization processes are saturated, sulfate buildup in the soil would occur. The rate of buildup will then at some point be accentuated by decreasing microbial sulfur requirements resulting from the effect of lower soil pH on microbial activity.

Table 4.11. Soil pH After 5 Years of Exposure to SO_2

	Soil pH		
	October 1979		May 1980
SO_2 Concentration	0–1 cm[1]	1–5 cm[1]	0–1 cm[1]
Control	5.59[a2]	5.35[a]	5.63[a]
Low	5.20[b]	5.23[a]	5.16[b]
Medium	5.25[b]	5.49[a]	5.11[b]
High	4.63[c]	5.35[a]	4.61[c]

[1]Soil depth.
[2]Values represent a mean of four and six determinations for each SO_2 treatment for 1979 and 1980, respectively, and each determination is a pool of four soil samples. Means not sharing a common superscript within a column are significantly different ($P < 0.05$).

Soil pH at the 0- to 1-cm depth, for the same samples that had been used for total sulfur determinations, was significantly lowered on both sample dates by the SO_2 treatments (Table 4.11). The 1980 samples were taken one half year after termination of the SO_2 treatments; therefore, it is not possible to establish a relationship between soil pH and time of exposure; the 2 years of data are presented for substantiation purposes only. No significant treatment effects were observed for soil pH at the 1- to 5-cm depth. The response patterns for soil pH and soil sulfate were essentially identical. Although this grassland displayed a high buffering capacity, the lowering of soil pH can affect the system by altering nutrient availabilities (Baker et al., 1977), decreasing N fixation (Rice et al., 1977) and microbial activity (Grant et al., 1979), thereby decreasing nutrient and carbon turnover.

4.4 Summary

The deposition and dynamics of sulfur in a terrestrial ecological system exposed to SO_2 concern the interface between the level and duration of exposure and the response of the various living components of the system. The concentrations and form of sulfur in various components of the system, when coupled with the responses to those concentrations, indicate the direct effects of particular concentrations of SO_2. Since the various components of a system are in a continual state of turnover and flux, the flow of sulfur after deposition and the accumulation of the sulfur in various sinks indicate indirect effects of particular concentrations of SO_2 and of the potential location and mode of long-term effects.

The aboveground portion of temperate grassland vegetation undergoes a near-complete turnover on an annual basis. Individual organs turn over within a single growing season. There appears to be very little redistribution of sulfur out of senescing leaves or carryover of SO_2–sulfur in the vegetation from one growing season to the next. Therefore, the direct effects of SO_2 on vegetation are in response to factors that influence the uptake of sulfur during a particular growing season. Aside from the concentration of SO_2 and the abiotic factors that influence uptake, we observed that differences in productivity among years, the nutrient status of the soil, and defoliation influence the sulfur concentration of aboveground herbage.

Large increases in the sulfur concentration of aboveground plants were not accompanied by increases in the sulfur concentration of belowground plant organs. We did, however, observe a disruption of the usual pattern in the transfer of sulfur between roots and shoots as evidenced by lower translocation rates and lower proportions of organic sulfur in belowground organs.

The absorption of SO_2–sulfur by vegegation was the predominant means of input of sulfur to the system. After death of the vegetation, the sulfur that had been accumulated during the growing season is incorporated into the litter pool. The residence time of sulfur in the litter is short. A large portion of the sulfur in new litter at the end of the growing season is released to the soil by the following spring.

The soil is the final sink for SO_2–sulfur. Therefore, long-term, cumulative, indirect effects of SO_2 exposure to all components of the system will be mediated through the effects of sulfur loading to the soil. One of the important effects of increased sulfur content of soil was the lowering of soil pH in the surface layer (1 cm). Our data show that SO_2 concentrations at the level of the high treatment for a duration of five growing seasons lowered surface soil pH 1 unit and depressed decomposition rates. These impacts are the same as those reported for wet deposition of SO_2–sulfur (Baker et al., 1977; Francis et al., 1980). Our field exposure system was designed to assess gaseous, dry deposition. Although comparisons of grossfall with throughfall (indicating rain intercepted by the plant canopy: Baker et al., 1977; Raybould et al., 1977) suggest that dry deposition is greater than wet deposition, the long-term cumulative effects of the two modes of deposition will be additive.

References

Ayazloo, M. J., N. B. Bell, and S. G. Garsed. 1980. Modification of chronic sulfur dioxide injury to *Lolium perenne* L. by different sulfur and nitrogen nutrient treatments. *Environ. Pollut.* (Ser. A) 22:295–307.

Baker, J. D. Hocking, and M. Nyborg. 1977. Acidity of open and intercepted precipitation in forests and effects on forest soils in Alberta, Canada. *Water Air Soil Pollut.* 7:449–460.

Banwart, W. L., and J. M. Bremner. 1976. Volitalization of sulfur from unamended and sulfate-treated soils. *Soil Biol. Biochem.* 8:19–22.

Barrow, N. J. 1975. Reactions of fertilizer sulfate in soils. In *Sulfur in Australasian Agriculture*, K. D. McLachlan, ed. pp. 50–57. Australia: Sydney University Press.

Biscoe, P. V., M. H. Unsworth, and H. R. Pinckney. 1973. The effects of low concentrations of sulfur dioxide on stomatal behavior in *Vicia faba*. *New Phytol.* 72:1299–1306.

Black, V. J., and M. H. Unsworth. 1980. Stomatal responses to sulfur dioxide and vapor pressure deficit. *J. Exp. Bot.* 31:667–677.

Bokhari, U. G., and J. S. Singh. 1975. Standing state and cycling of nitrogen in soil-vegetation components of prairie ecosystems. *Ann. Bot.* 39:273–285.

Bouchard, R., and H. R. Conrad. 1974. Sulfur metabolism and nutritional changes in lactating cows associated with supplemental sulfate and methionine hydroxy analog. *Can. J. Anim. Sci.* 54:587–593.

Bouma, D. 1975. The uptake and translocation of sulfur in plants. In *Sulfur in Australasian Agriculture*, K. D. McLachlan, ed. pp. 79–86. Australia: Sidney University Press.

Brady, C. J. 1973. Changes accompanying growth and senescence and effect of physiological stress. In *Chemistry and Biochemistry of Herbage*, G. W. Butler and R. W. Bailey, eds. Vol. 2, pp. 317–351. New York: Academic Press.

Brouwer, R. 1963. Some aspects of the equilibrium between overground and underground plant parts. *Inst. Biol. Scheikd. Anderz. Landbouwgenwassen Wageningen Jaarb.* 1963:31–39.

Buckman, H. O., and N. C. Brady. 1969. *The Nature and Properties of Soils*. 7th ed. London: The Macmillan Company, Collier-Macmillan Limited.

Caput, C., U. Belot, D. Auclair, and N. Decourt. 1978. Absorption of sulfur dioxide by pine needles leading to acute injury. *Environ. Pollut.* 16:3–16.

Carlson, R. W., and F. A. Bazzaz. 1982. Photosynthetic and growth response to fumigation with SO_2 at elevated CO_2 for C_3 and C_4 plants. *Oecologia (Berl.)* 54:50–54.

Compton, O. C., and C. F. Remmert. 1960. Effects of airborne fluorine on injury and fluorine content of gladiolus leaves. *Proc. Am. Soc. Hort. Soc.* 75:663–675.

Coughenour, M. B., J. L. Dodd, D. C. Coleman, and W. K. Lauenroth. 1979. Partitioning of carbon and SO_2 sulfur in a native grassland. *Oecologia (Berl.)* 42:229–240.

Cowling, D. W., and A. W. Bristow. 1979. Effects of SO_2 on sulfur and nitrogen fractions and on free amino acids in perennial ryegrass. *J. Sci. Fd. Agric.* 30:354–360.

Cowling, D. W., and D. R. Lockyer. 1978. Effect of SO_2 on *Lolium perenne* grown at different levels of sulfur and nitrogen. *J. Exp. Bot.* 29:257–265.

Davis, D. D., and R. G. Wilhour. 1976. Susceptibility of woody plants to sulfur dioxide and photochemical oxidants. Washington, D.C.: EPA Ecological Research Series. EPA-600/3-76-102.

de Cormis, L., J. Cantuel, and J. Bonte. 1969. Contribution à l'étude de l'absorption du soufre par les plantes soumises à une atmosphère contenant du dioxyde de soufre. *Pollut. Atmos. (Paris)* 44:195–202.

dela Fuente, R. K., and A. C. Leopold. 1968. Senescence processes in leaf abscission. *Plant Physiol.* 43:1496–1502.

Dijkshoorn, W. 1959. The rate of uptake of chloride, phosphate and sulfate in perennial ryegrass. *Neth. J. Agr. Sci.* 7:194–201.

Dijkshoorn, W., and A. L. Van Wijk. 1967. The sulfur requirements of plants as evidenced by the sulfur-nitrogen ratio in the organic matter. A review of published data. *Plant Soil* 26:129–157.

Dodd, J. L., and W. K. Lauenroth. 1981. Effects of low-level SO_2 fumigation on decomposition of western wheatgrass litter in a mixed-grass prairie. *Water Air Soil Pollut.* 15:257–261.

Eriksson, E. 1963. The yearly circulation of sulfur in nature. *J. Geophys. Res.* 68:4001–4008.

Faller, N. 1971. Plant nutrient sulphur—SO_2 vs. SO_4. *Sulfur Inst. J.* 7:5–6.

Fitzgerald, J. W. 1978. Naturally occurring organosulfur compounds in soils. In *Sulfur in the Environment, Part II*, J. O. Nriagu, ed. pp. 391–443. New York: Wiley.

Francis, A. J., D. Olson, and R. Bernatsky. 1980. Effect of acidity on microbial processes in a forest soil. In *Ecological Impact of Acid Precipitation*, D. Drabløs and A. Tollan, eds. pp. 166–167. Norway: Norwegian Interdisciplinary Research Programme (SNSF), As-NLH.

Freid, M. 1948. The absorption of sulfur dioxide by plants as shown by the use of radioactive sulfur. *Soil Sci. Soc. Am. Proc.* 13:135–138.

Freney, J. R., and R. J. Swaby. 1975. Sulfur transformations in soils. In *Sulfur in Australasian Agriculture*, K. D. McLachlan, ed. pp. 31–39. Australia: Sydney University Press.

Garsed, S. G., and D. J. Read. 1977. The uptake and matebolism of $^{35}SO_2$ in plants of differing sensitivity to sulfur dioxide. *Environ. Pollut.* 13:173–187.

Garsed, S. G., A. J. Rutter, and J. Relton. 1981. The effects of sulfur dioxide on the growth of *Pinus sylvestris* on two soils. *Environ. Pollut.* (Series A) 24:219–232.

Godzik, S., and H. F. Linskens. 1974. Concentration changes of free amino acids in primary bean leaves after continuous and interrupted SO_2 fumigation and recovery. *Environ. Pollut.* 7:25–38.

Grant, I. F., K. Bancroft, and M. Alexander. 1979. SO_2 and NO_2 effects on microbial activity in an acid forest soil. *Microbial Ecol.* 5:85–89.

Guderian, R. 1977. *Air pollution. Phytotoxicity of Acidic Gases and Its Significance in Air Pollution Control.* Berlin: Springer-Verlag Ecological Studies 22.

Hanson, E. A., B. S. Barrien, and J. G. Wood. 1941. Relations between protein-nitrogen, protein-sulphur and chlorophyll in leaves of sudan grass. *Aust. J. Exp. Biol. Med. Sci.* 19:231–234.

Harborne, J. B. 1977. *Introduction to Ecological Biochemistry*. London: Academic Press.

Hill, A. C. 1969. Air quality standards for fluoride vegetation effects. *J. Air Pollut. Control Assoc.* 19:331–336.

Hocking, D., and M. B. Hocking. 1977. Equilibrium solubility of trace atmospheric sulfur dioxide in water and its bearing on air pollution injury to plants. *Environ. Pollut.* 13:57–64.

Hubbert, F., Jr., E. Cheng, and W. Burroughs. 1958. Mineral requirements of rumen microorganisms for cellulose digestion *in vitro*. *J. Anim. Sci.* 17:559–568.

Jäger, H. J. 1975. Effect of sulfur dioxide fumigation on the activity of enzymes of the amino acid metabolism and the free amino acid contents in plants of varying resistance. *Z. Planzenkr. Pfanzenschutz*. 82:139–148.

Jäger, H. J., and H. Klein. 1977. Biochemical and physiological detection of sulfur dioxide injury to pea plants (*Pisium sativum*). *Air Pollut. Control Assoc. J.* 27:464–465.

Jensen, K. F., and T. T. Kozlowski. 1975. Absorption and translocation of sulfur dioxide by seedlings of four forest tree species. *J. Environ. Qual.* 4:379–382.

Klein, H., H. J. Jäger, W. Domes, and C. H. Wong. 1978. Mechanisms contributing to differential sensitivities of plants to sulfur dioxide. *Oecologia* (*Berl.*) 33:203–208.

Kylin, A. 1960. The incorporation of radio-sulfur from external sulfate into different sulfur fractions of isolated leaves. *Physiologia Pl.* 13:355–379.

Lauenroth, W. K., C. J. Bicak, and J. L. Dodd. 1979. Sulfur accumulation in western wheatgrass exposed to three controlled SO_2 concentrations. *Plant Soil* 53:131–136.

Lauenroth, W. K., J. K. Detling, and J. L. Dodd. 1983. Impact of SO_2 exposure on the response of *Agropyron smithii* to defoliation. (Submitted for publication).

Leetham, J. W., J. L. Dodd, and W. K. Lauenroth. 1983. Effects of low-level sulfur dioxide exposure on decomposition of *Agropyron smithii* litter under laboratory conditions. *Water Air Soil Pollut.* 19:247–250.

Leibholz, J., and R. W. Naylor. 1971. The effect of urea in the diet of the early-weaned calf on weight gain, nitrogen and sulfur balance, and plasma urea and free amino acid concentrations. *Aust. J. Agric. Res.* 22:655–662.

Leininger, W. C., and J. E. Taylor. 1981. Germination and seedling establishment as affected by sulfur dioxide. Section 13. In *The Bioenvironmental Impact of a Coal-Fired Power Plant*, E. M. Preston, D. W. O'Guinn, and R. A. Wilson, eds. Sixth Interim Report, Colstrip, Montana. Corvallis, Oregon: Environmental Protection Agency.

Lockyer, D. R., D. W. Cowling, and J. S. Fenton. 1978. Laboratory measurements of dry deposition of sulfur dioxide onto several soils from England and Wales. *J. Sci. Fd. Agric.* 29:739–746.

Majernik, O., and T. A. Mansfield. 1970. Direct effect of sulfur dioxide pollution on the degree of opening of stomata. *Nature* (*London*) 227:377–378.

Majernik, O., and R. A. Mansfield. 1972. Stomatal responses to raised atmospheric CO_2 concentrations during exposure of plants to SO_2 pollution. *Environ. Pollut.* 3:1–7.

Malhotra, S. S., and S. K. Sarkar. 1979. Effects of sulfur dioxide on sugar and free amino acid content of pine seedlings. *Physiol. Plant.* 47:223–228.

Metson, A. J. 1973. Sulfur in forage crops. Plant analysis as a guide to the sulfur status of forage grasses and legumes. The Sulfur Institute. Tech. Bull. No. 20.

Milchunas, D. G., W. K. Lauenroth, and J. L. Dodd. 1981a. Forage quality of western wheatgrass exposed to sulfur dioxide. *J. Range Manage.* 34:282–285.

Milchunas, D. G., W. K. Lauenroth, and J. L. Dodd. 1981b. The effect of SO_2 on ^{14}C translocation in *Agropyron smithii* Rydb. *J. Environ. Exp. Bot.* 22:81–91.

Milchunas, D. G., W. K. Lauenroth, and J. L. Dodd. 1983. The interaction of atmospheric and soil sulfur on the sulfur and selenium ceoncentration of range plants. *Plant Soil* 72:117–125.

Mishra, L. C. 1980. Effects of sulfur dioxide fumigation on groundnut (*Arachis hypogaea* L.). *Environ. Exp. Bot.* 20:397–400.

Moir, R. J., M. Somers, and A. C. Bray. 1967. Utilization of dietary sulfur and nitrogen. *Sulfur Inst. J.* 3:15–18.

Nyborg, M. 1978. Sulfur pollution and soils. p. 359–390. In *Sulfur in the Environment, Part II: Ecological Impacts*, J. O. Nriagu, ed. New York: Wiley.

Nyborg, M., J. Crepin, D. Hocking, and J. Baker. 1977. Effect of sulfur dioxide on precipitation and on the sulfur content and acidity of soils in Alberta, Canada. *Water Air Soil Pollut.* 7:439–448.

Odum, E. P., J. T. Finn, and E. H. Franz. 1979. Perturbation theory and the subsidy-stress gradient. *BioScience* 29:349–352.

Pate, J. S. 1965. Roots as organs of assimilation of sulfate. *Science* 149:547–548.

Payrissat, M., and S. Beilke. 1975. Laboratory measurements of the uptake of sulfur dioxide by different European soils. *Atmos. Environ.* 9:211–217.

Pitman, M. G. 1976. Ion uptake by plant roots. In *Transport in Plants II: Part B, Tissues and Organs*, U. Lüttage and M. G. Pitman, eds. pp. 95–128. Berlin: Springer-Verlag.

Raybould, C. C., M. H. Unsworth, and P. J. Gregory. 1977. Sources of sulfur in rain collected below a wheat canopy. *Nature* (*London*) 267:146–147.

Redmann, R. E., and Z. M. Abouguendia. 1978. Partitioning of respiration from soil, litter and plants in a mixed-grassland ecosystem. *Oecologia* (*Berl.*) 36:69–79.

Rennenberg, H., R. Schmitz, and L. Bergmann. 1979. Long-distance transport of sulfur in *Nicotiana tabacum*. *Planta* 147:57–62.

Rice, W. A., D. C. Penney, and M. Nyborg. 1977. Effects of soil acidity on rhizobia numbers, nodulation and nitrogen fixation by alfalfa and red clover. *Can. J. Soil Sci.* 57:197–203.

Rice, P. M., L. H. Pye, R. Boldi, J. O'Loughlin, P. C. Tourangeau, and C. C. Gordon. 1979. The effects of "low level SO_2" exposure on sulfur accumulation and various plant life responses of some major grassland species on the ZAPS sites. In *Bioenvironmental Impact of a Coal-Fired Power Plant*, E. M. Preston and T. L. Gullett, eds. pp. 494–591. Corvallis, Oregon: U.S. Environmental Protection Agency, Corvallis Environmental Research Laboratory.

Salisbury, F., and C. Ross. 1969. *Plant Physiology*. Belmont, California: Wadsworth Publ. Co., Inc.

Schiff, J. A., and R. C. Hodson. 1973. The metabolism of sulfate. *Annu. Rev. Plant Physiol.* 24:381–414.

Scott, P. C., and A. C. Leopold. 1966. Abscission as a mobilization phenomenon. *Plant Physiol.* 11:826–830.

Smith, K. A., J. M. Bremner, and M. A. Tabatabai. 1973. Sorption of gaseous atmospheric pollutants by soils. *Soil Sci.* 116:313–319.

Thomas, M. D., and G. R. Hill. 1937. Relation of sulphur dioxide in the atmosphere to photosynthesis and respiration of alfalfa. *Plant Physiol.* 12:309–383.

Torii, K., and A. Fujiwara. 1967. Physiology of sulfate in higher plants II. Uptake and translocation of radioactive sulfate in sulfur deficient plants. *Tohoku J. Agr. Res.* 17:89–99.

Trinkle, A., E. Cheng, and W. Burroughs. 1958. Availability of different sulfur sources for rumen microorganisms in *in vitro* cellulose digestion. *J. Anim. Sci.* 17:1191 (Abstr.).

Tukey, H. B., Jr. 1970. The leaching of substances from plants. *Annu. Rev. Plant Physiol.* 21:305–324.

Unsworth, M. H., P. V. Biscoe, and H. R. Pinckney. 1972. Stomatal responses to sulfur dioxide. *Nature* (*London*) 239:458–459.

Wellburn, A. R., T. M. Capron, H. S. Chan, and D. C. Horsman. 1976. Biochemical effects of atmospheric pollutants on plants. In *Effects of Air Pollutants on Plants*, T. A. Mansfield, ed. pp. 105–114. London: Cambridge Univ. Press.

Whanger, P. D., P. H. Weswig, and J. C. Oldfield. 1978. Selenium, sulfur and nitrogen levels in rumen microorganisms. *J. Anim. Sci.* 46:515–519.

Williams, R. F. 1955. Redistribution of mineral elements during development. *Annu. Rev. Plant Physiol.* 6:25–42.

Winner, W. E., and H. A. Mooney. 1980. Ecology of SO_2 resistance: II. Photosynthetic changes of shrubs in relation to SO_2 absorption and stomatal behavior. *Oecologia* (*Berl.*) 44:296–302.

Witkamp, M., and J. van der Drift. 1961. Breakdown of forest litter in relation to environmental factors. *Plant. Soil* 15:295–311.

Wood, J. G., and B. S. Barrien. 1939. Studies on the sulfur metabolism of plants. III. On changes in amounts of protein sulfur and sulfate sulfur during starvation. *New Phytol.* 38:265–272.

Ziegler, I. 1972. The effect of SO_2 on the activity of ribulose-1,5-diphosphate carboxylase in isolated spinach chloroplasts. *Planta* (*Berl.*) 103:155–163.

Ziegler, I. 1975. The effect of SO_2 pollution on plant metabolism. *Residue Rev.* 56:79–105.

5. Responses of the Vegetation

W. K. LAUENROTH, D. G. MILCHUNAS, AND J. L. DODD

5.1 Introduction

Sulfur dioxide, because of its potential as a nutrient source as well as a toxic agent, influences plants and plant communities in a variety of ways. Separating the nutrient and toxic effects is difficult, because a process or structure that is impacted positively by the nutrient-supplying properties of SO_2 may be negatively impacted by the same concentration of SO_2 as exposure time is increased. A transition from positive to negative effects can also be expected over a range of SO_2 concentrations for the same exposure time. Odum et al. (1979) described this general phenomenon for perturbed ecological systems as a subsidy–stress gradient.

In this chapter we report the results of our experimental investigations of the impacts of SO_2 on grassland plants at levels of organization from chlorophyll molecules to whole plant communities. We only included the material that was required to establish the main responses. More detailed information can be found in the literature articles and reports cited. The chapter ends with a summary of the most important results and a discussion of their ecological significance.

5.2 Canopy Structure

5.2.1 Leaf Area

The SO_2 treatments altered the proportion of leaf area in various strata of the canopy, as well as the height of the canopy. The low-SO_2 treatment displayed maximum canopy height, live leaf canopy profile fullness, total leaf area, and live to dead ratios. Plants on the medium- and high-SO_2 treatments never attained the leaf area observed on the control. The growth patterns indicated a positive response with low-SO_2 exposure, but negative responses predominated with increasing concentrations of SO_2.

Leaf area profiles for *A. smithii* illustrate the structure of the canopy at several times during a growing season (Figure 5.1). At the beginning of the growing-season, tillers were short with only three to four leaves. At this time no effects of SO_2 exposure were detectable. Sulfur dioxide effects on canopy structure were first evident in mid-June, which represents the average date of maximum live aboveground biomass in these grasslands. Although live leaf areas were greatest on the control, tillers on the low-SO_2 treatment were taller and had more leaves. As the growing season progressed, the number of leaves and the height of tillers remained greatest on the low-SO_2 treatment. Maximum current season biomass is typically measured in July in these grasslands. At this time, the low-SO_2 treatment had the greatest live leaf area throughout the profile. The control, medium-, and high-SO_2 treatments had similar live leaf areas, although the distributions were greater in the lower strata on the SO_2 treatments and greater in the upper strata on the control. September usually marks the end of the major growth period for the year although, depending upon precipitation, a secondary growth period may occur in September and October. By September, live leaf area on control tillers was found only in the upper canopy layer, the low-SO_2 treatment still had live leaves in the middle of the canopy as well as at the top, and the medium- and high-SO_2 treatments were intermediate between the low-SO_2 treatment and the control. The profile of dead leaf area in September indicated more dead leaves on the control. The SO_2 treatments had similar profiles of dead leaf area.

To further assess the balance between positive and negative effects of SO_2 on the vegetation we designed an experiment at Site I to evaluate the interactions of SO_2 exposure with nitrogen and sulfur fertilization (Milchunas et al., 1981b; Lauenroth et al., 1983). The addition of nitrogen fertilizer in combination with the SO_2 treatments had a substantial impact upon canopy structure. Total live and dead leaf areas were increased, live to dead leaf area ratios were altered, and leaf area profiles were affected (Figure 5.2). The largest impact of nitrogen fertilizer on leaf areas was observed for the low-$SO_2 \times N$ treatment. In all cases the interaction of nitrogen fertilization with SO_2 exposure increased leaf area compared to the control. Nitrogen fertilizer increased the live/dead leaf area ratios in June and either decreased or had no influence on the ratios in July or September.

5.2.2 Chlorophyll Concentration

Light interception increases with increasing leaf area in a system such as a mixed prairie where shading effects are minimal. However, increased light interception may or may not increase the photosynthetic capacity of a canopy exposed to SO_2. Sulfur dioxide exposure can decrease the chlorophyll content in lichens (Rao and LeBlanc, 1965), bryophytes (Coker, 1967), and pine needles

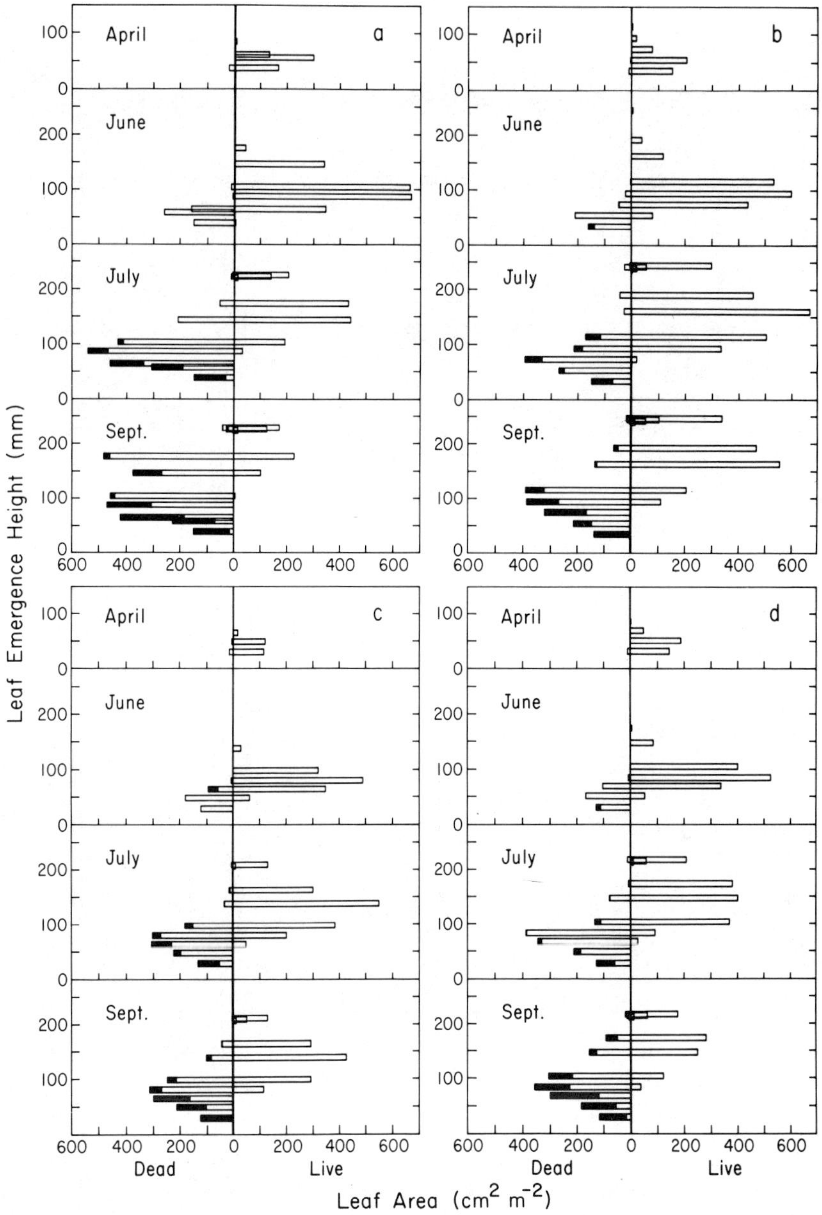

Figure 5.1. Canopy profiles of leaf area ($cm^2 \cdot m^{-2}$ ground area) for *Agropyron smithii* on the control (a), low- (b), medium- (c), and high- (d) concentration SO_2 treatments, Site I, 1978. The shaded portions of the dead leaf area bars repesent loss to litter. When a previous measurement of a particular leaf was greater than the current measurement, the negative value represented loss to litter.

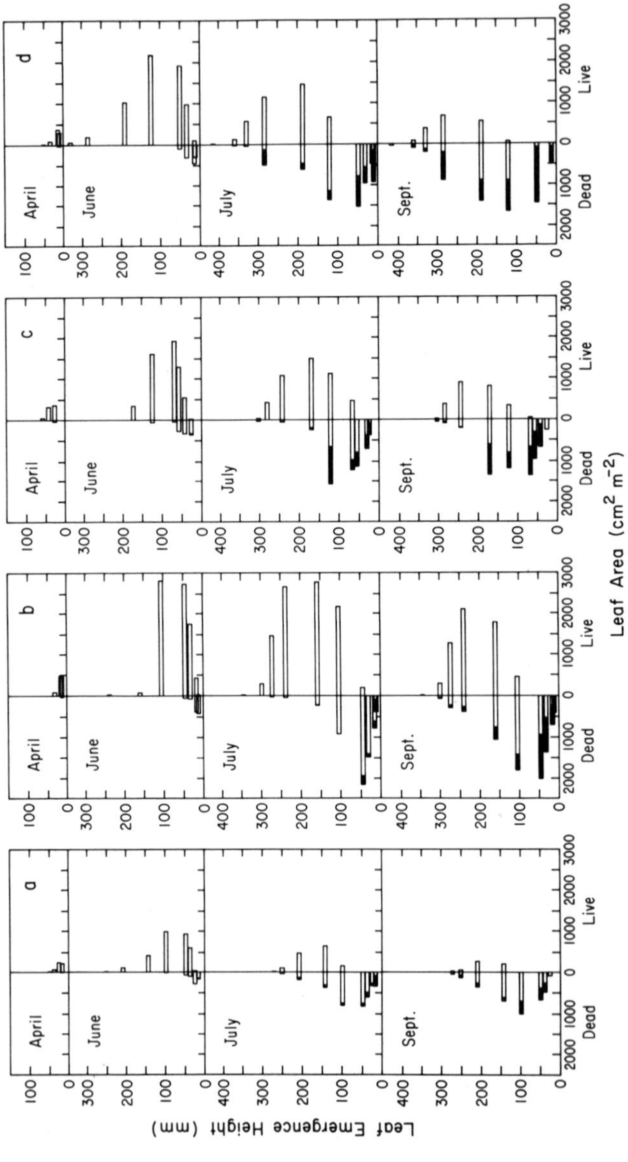

Figure 5.2. Canopy profiles of leaf area (cm$^2 \cdot$ m^{-2} ground area) for *Agropyron smithii* on the control (a), low- (b), medium- (c), and high- (d) SO_2 treatments with nitrogen fertilizer, Site I, 1978. The shaded portion of the dead leaf area bars represent loss to litter.

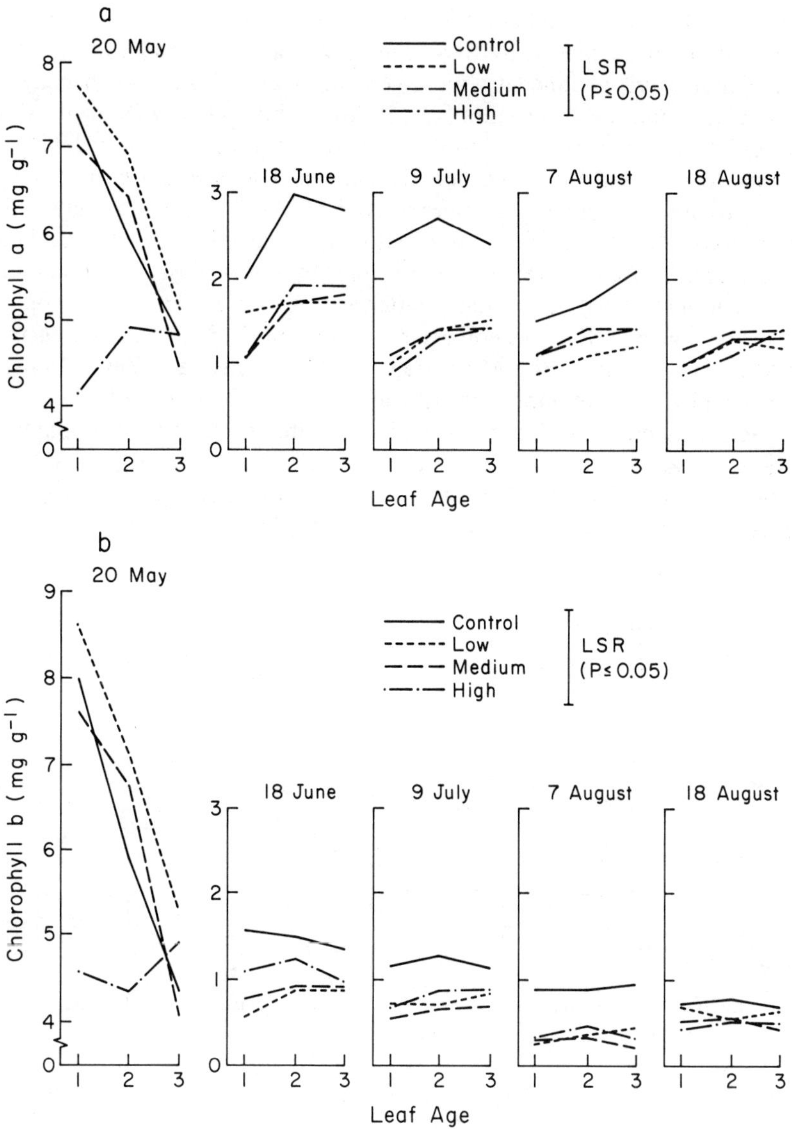

Figure 5.3. Chorophyll *a* (a) and *b* (b) concentration ($mg \cdot g^{-1}$) of *Agropyron smithii* leaves of three different ages (1 = oldest leaf) exposed to three controlled SO_2 treatments. (Adapted from Lauenroth and Dodd, 1981a.)

(Malhotra, 1977). This decrease in chlorophyll content with SO_2 exposure is a result of the conversion of chlorophyll to phaeophytin (Malhotra, 1977). Although chlorophyll content may not be the only important explanatory variable for differences in photosynthetic rates among species (Hesketh, 1963), one should expect positive relationships between leaf chlorophyll content and photosynthesis (Reutz, 1973; Buttery and Buzzel, 1977).

Analysis of *A. smithii* leaf blades for chlorophyll *a* and *b* concentrations showed significant differences caused by SO_2, date, and leaf age (Lauenroth and Dodd, 1981a). Chlorophyll *a* concentrations were highest in May and then declined through the growing season (Figure 5.3). In May, chlorophyll *a* concentrations in the two oldest leaves were lower on the high-SO_2 treatment compared with other treatments. In June and July, chlorophyll *a* concentrations were similar among the three SO_2 treatments, but lower than the control. As the season progressed, the concentrations of chlorophyll *a* on the control approached the levels found on the SO_2 treatments and all treatments were similar by the August 18 sampling date. Similar treatment × leaf age × date patterns were observed for chlorophyll *b* except the response to SO_2 was not as large, with significant treatment effects only on the May date (Figure 5.3b). Malhotra (1977) and Peiser and Yang (1977) also found chlorophyll *a* to be more sensitive to SO_2 than chlorophyll *b*.

The SO_2 treatments did not have similar effects on the chlorophyll concentrations of a variety of plant species (Lauenroth and Dodd, 1981b). The responses observed for six of eight species examined ranged from significant increases to significant decreases in chlorophylls *a* and *b* (Figure 5.4). The most

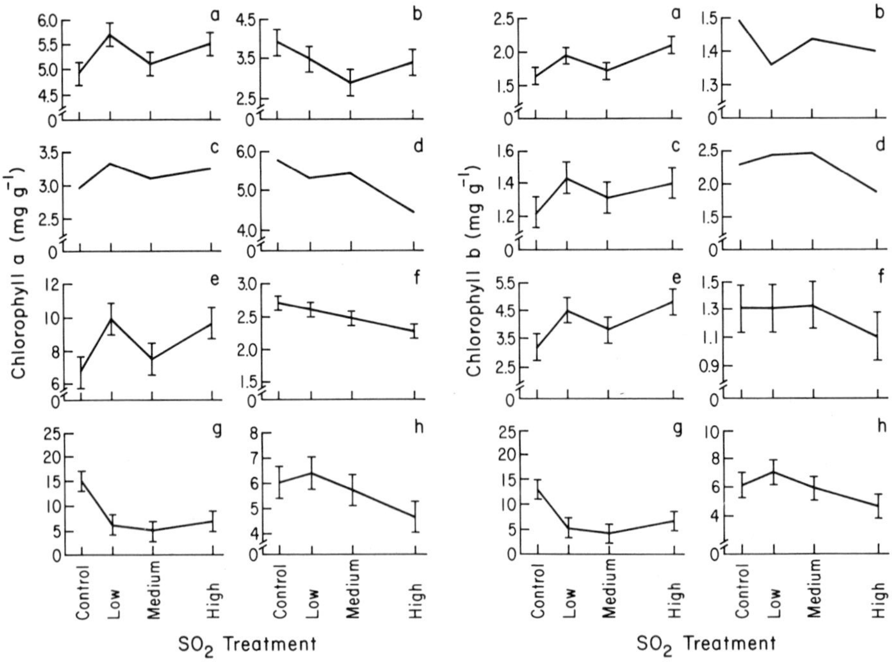

Figure 5.4. Responses of chlorophyll *a* (a) and chlorophyll *b* (b) concentrations ($mg \cdot g^{-1}$) to three concentrations of SO_2 for (a) *Psoralea argophylla*; (b) *Sphaeralcea coccinea*; (c) *Tragopogon dubius*; (d) *Koeleria cristata*; (e) *Taraxacum officinale*; (f) *Achillea millefolium*; (g) *Bromus japonicus*; and (h) *Agropyron smithii*. Vertical bars represent least significant ranges ($P \leq 0.05$). No bars indicate no significant differences. (Adapted from Lauenroth and Dodd, 1981b.)

sensitive species sampled was *Bromus japonicus*, with both chlorophylls *a* and *b* significantly reduced by the presence of SO_2. Chlorophyll *a* concentrations of *Achillea millefolium* decreased with increasing SO_2 concentration.

Chlorophyll concentration is often cited as a sensitive and reliable indicator of SO_2 exposure (Heck et al., 1979; Knabe, 1976; Linzon, 1978). Our data showed that caution must be exercised in applying this generality. Individual species respond differently, plant phenological development influences the response, and large variability should be expected.

5.2.3 Cover

Canopy cover expressed as the sum of the cover for each species (Figure 5.5) (values greater than 100% indicate overlap among species) ranged between 80 and 130% (Taylor et al., 1980). The lowest values were associated with the first growing season after excluding domestic livestock and the dry 1977 growing season. The general trend between 1975 and 1978 was an increase in cover as a result of exclusion of cattle on all plots regardless of SO_2 concentration. The largest increase was recorded for the controls. Whether this was an impact of SO_2 on the remaining plots was not clear.

The contribution of *Agropyron smithii* to total canopy cover ranged from 15 to 25% on Site I and from 20 to 45% on Site II (Figure 5.6). On Site I, cover changes for *A. smithii* were small and similar in direction and degree of variability to those for total cover. On Site II, there was a decrease in the cover of *A. smithii* on the medium-SO_2 treatment. This result is supported by tiller density measurements for both 1977 and 1978. In neither case was the decrease statistically significant. Dodd et al. (1982) did not find any significant impacts of SO_2 on *A. smithii* biomass production.

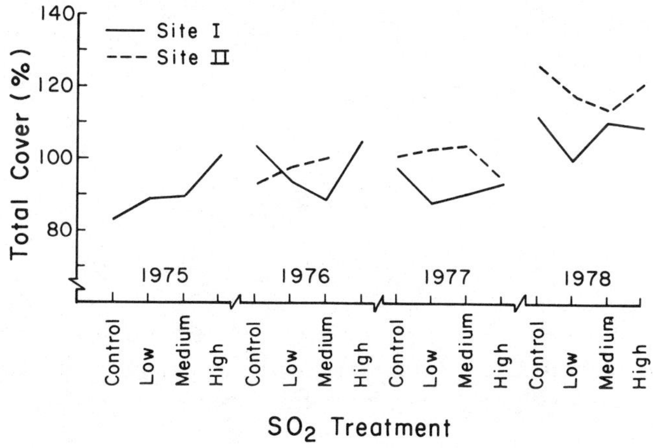

Figure 5.5. Canopy cover (%) for the total live vegetation on Sites I and II, 1975–1978. (Adapted from Taylor et al., 1980.)

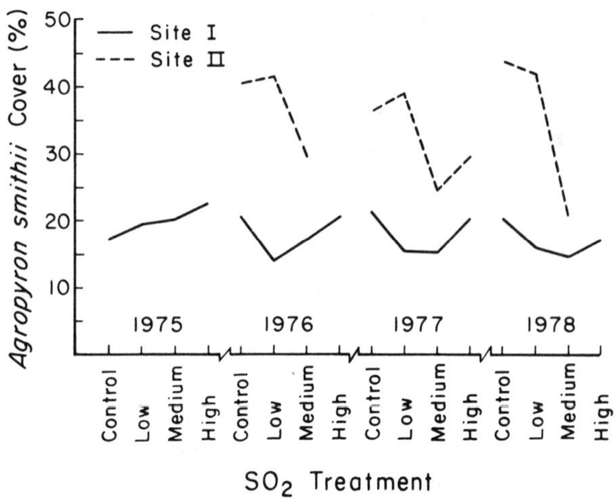

Figure 5.6. Canopy cover (%) for *Agropyron smithii* on Sites I and II, 1975–1978. (Adapted from Taylor et al., 1980.)

The percentage of the soil surface covered by plant litter increased substantially over the experiment as a result of protection from grazing by cattle. No clear impacts of SO_2 were observed despite reduced litter disappearance (Table 4.9). Litter biomass showed corroborating increases (see Figure 5.9).

Lichens are more sensitive to SO_2 exposure than higher plants and have therefore often been used as bioindicators of air pollution (Hanksworth and Rose, 1970; Skye, 1968). Their sensitivity can partly be explained by their ability to accumulate nutrients from very dilute solutions. Lichens lack cuticules and stomates, therefore gas-exchange and nutrient uptake occur over the entire surface. Eversman (1978) reported plasmolysis of algal cells in lichens suspended aboveground in the SO_2 plots but no plasmolysis for ground-level samples. The relative cover of lichens decreased on the SO_2 treatments over the years of exposure (Figure 5.7).

5.3 Biomass

Perhaps better than any other single variable, the distribution of plant biomass defines the structure of terrestrial ecological systems. A characteristic of grasslands, particularly those in semiarid regions, is a distribution of biomass to which grasses contribute the largest amount (Singh et al., 1983) and in which the amount occurring belowground is substantially greater than that aboveground (Sims et al., 1978).

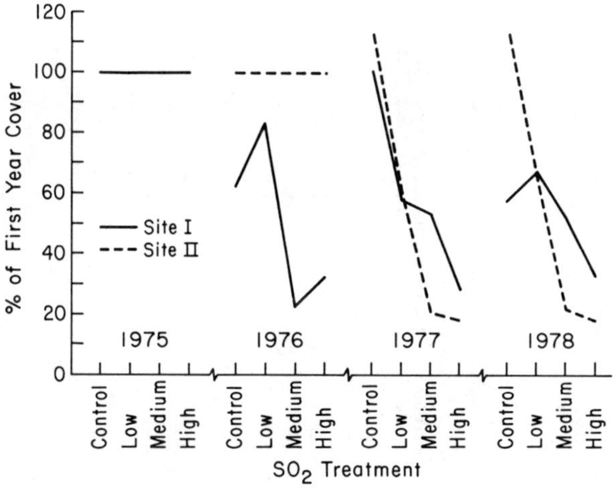

Figure 5.7. Lichen cover expressed as a percentage of first year cover, Site I and II. 1975–1978. (Adapted from Taylor et al., 1980.)

Aboveground biomass, particularly total standing crop and live standing crop, is very dependent upon current biomass production. While biomass was greater in 1975 than 1978 (Table 5.1), no significant effects resulting from SO_2 exposure were detectable. Cool-season grasses clearly remained the most important functional group on all treatments (Figure 5.8). Cool-season grass biomass ranged from 60 to 90% of total July standing crop. The most important species in this

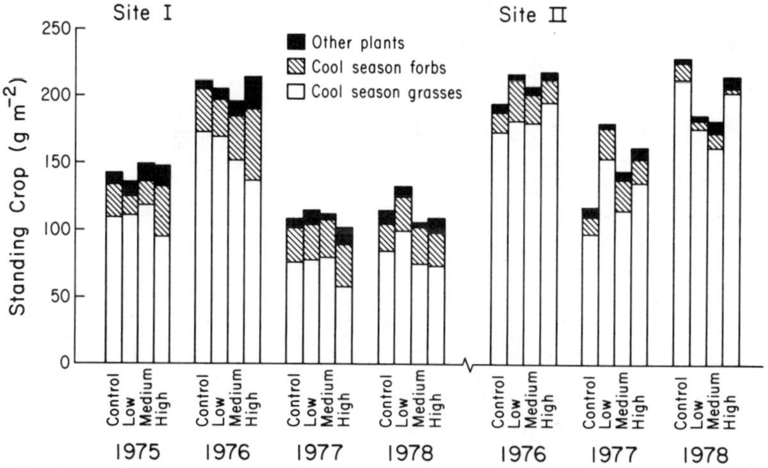

Figure 5.8. Seasonal average biomass (g · m^{-2}) of plant functional groups, 1975–1978, Sites I and II.

Table 5.1. July Standing Crop Biomass ($g \cdot m^{-2}$) Measured on Site in 1975 and 1978

	Total Standing Crop	Live Standing Crop	Litter	Roots[a] (0–60 cm)	Crowns	Rhizomes	Belowground/Aboveground
1975							
Control	139	126	123	959	53	26	2.7
Low	136	125	155	684	55	32	1.9
Medium	150	139	155	886	43	24	2.1
High	148	130	154	651	33	24	1.6
1978							
Control	115	105	221	998	94	44	2.6
Low	132	129	254	912	115	33	2.1
Medium	104	103	269	947	87	32	2.2
High	109	106	271	893	80	26	2.1

[a]Root biomass in 1978 was calculated by roots (0–60 cm) = roots (0–10 cm)/0.55.

Table 5.2. Effect of Three Levels of SO_2 Exposure on Average July Standing Crop ($g \cdot m^{-2}$) of Current-Season Biomass for Sites I and II[1]

	Site I				Site II			
	Control	Low	Med.	High	Control	Low	Med.	High
Cool-season grasses (CSG)								
Agropyron smithii	60	49	48	55	95	124	79	105
Bromus japonicus	4	1	1	<1	30	14	8	3
Koeleria cristata	25	30	29	19	5	2	22	32
Poa spp.	2	4	4	3	12	19	23	22
Stipa comata	5	11	7	1	<1	<1	<1	<1
Stipa viridula	<1	<1	<1	<1	6	2	10	10
Other CSG	3	7	8	1	<1	1	<1	<1
Total[2]	94	101	96	79	117	148	134	168
Cool-season forbs (CSF)								
Achillea millifolium	8	2	4	17	1	6	2	5
Tragopogon dubius	10	13	12	7	5	1	5	4
Taraxacum officinale	2	1	4	6	4	12	8	3
Other CFS	5	7	7	6	3	1	3	1
Total	25	23	27	36	14	20	19	13
Other plants	6	6	6	13	3	2	6	6

[1]Adapted from Dodd et al. (1982).
[2]This does not include *B. japonicus*.

group, *A. smithii* and *K. cristata*, showed no response to SO_2 treatments on either site (Dodd et al., 1982) (Table 5.2); together they accounted for 80 to 90% of the total cool-season biomass. Cool-season forbs were the only other consistently important group and contributed 10 to 30% of July standing crop.

The only species that showed a relationship with SO_2 exposure was the winter annual grass *Bromus japonicus*. July standing crop for *B. japonicus* was significantly reduced by SO_2 (Table 5.2; see also Figure 5.24). Chlorophyll concentrations were also significantly reduced in *Bromus japonicus* as a result of SO_2 exposure (see Figure 5.4).

Litter represents a large storage pool of carbon and mineral nutrients in the system. The only substantial pattern evident over the period of the study was an increase in litter biomass, which was related to protection from grazing by domestic cattle (Figure 5.9). The increase was generally linear for both sites. Despite the observed reduction in litter disappearance in mesh bags as a result of SO_2 (see Table 4.9, Section 4.3.5), we found no effects of SO_2 on either cover of litter (Taylor et al., 1980) or litter biomass.

Our results for the effects of SO_2 exposure on belowground plant organs indicated no consistent measurable impact on crowns, rhizomes, or roots (Table 5.3). Year effects attributable to exclusion of large herbivores were the only

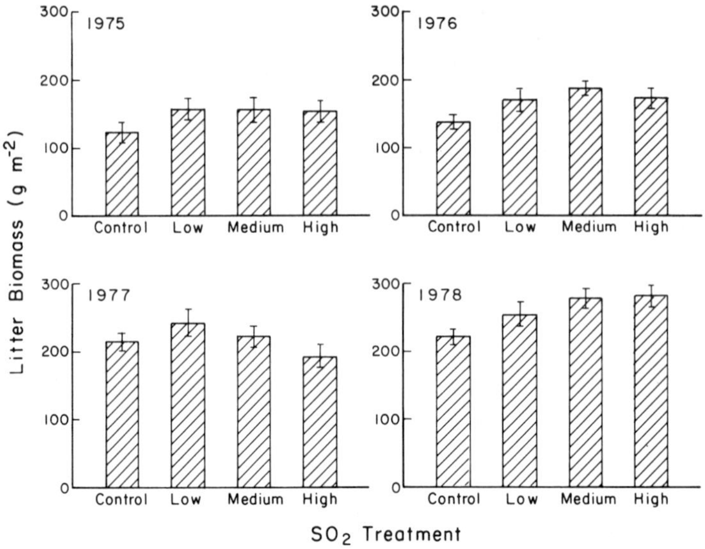

Figure 5.9. Litter biomass (g · m^{-2}) at the time of peak standing herbage biomass, Site I, 1975–1978. Values are on an ash-free, dry weight basis.

consistent significant effects observed. The distribution of belowground biomass with depth in the soil was investigated only during 1975 and 1976 and was not influenced by exposure to SO_2.

Table 5.3. Average Standing Crop (g · m^{-2}) of Belowground Biomass by Morphological Categories (0- to 10-cm depth)[1]

	Year			SO_2 Treatment			
Site I	1975	1976	1978	Control	Low	Medium	High
Crown	51[a]	70[b]	97[c]	74[a]	77[a]	74[a]	66[a]
Rhizome	25[a]	32[b]	30[b]	37[b]	26[a]	27[a]	26[a]
Roots	549[a]	528[a]	677[b]	586[a]	573[a]	582[a]	596[a]
Total	624[a]	629[a]	803[a]	697[a]	676[a]	682[a]	687[a]

	Year		SO_2 Treatment			
Site II	1976	1978	Control	Low	Medium	High
Crown	71[a]	99[b]	79[a]	88[a]	84[a]	88[a]
Rhizome	28[a]	32[b]	26[b]	35[b]	31[ab]	29[a]
Roots	593[a]	682[b]	593[a]	669[a]	661[a]	626[a]
Total	692[a]	813[b]	697[a]	791[a]	775[a]	744[a]

[1]Means for each site within a row for each year or treatment not followed by the same letter are significantly different ($P \leqq 0.05$). Adapted from Dodd et al. (1982).

5.4 Carbon Allocation

Partitioning of photoassimilated carbon among various plant organs, and between shoots and roots, is an important aspect in the functioning of a plant as an integrated system. Translocation processes and the maintenance of carbon balance are subject to regulation involving both positive and negative feedback mechanisms (Geiger, 1979). Export of carbon from a source leaf is a function of the availability of sucrose and other mobile molecules and the demand for these assimilates in other locations. Sulfur dioxide exposure has been shown to affect photosynthesis, plant growth, and carbohydrate metabolism in plants (Ziegler, 1975). Short-term exposure to SO_2 has been reported to inhibit phloem carbon translocation (Teh and Swanson, 1979).

We conducted two experiments to investigate the translocation of ^{14}C in plants growing on the control and high-SO_2 treatments. The first experiment, conducted in 1976, involved the exposure of the vegetation under a 0.5-m^2 ground area tent to both $^{35}SO_2$ and $^{14}CO_2$ (Coughenour et al., 1979). This study addressed carbon and sulfur translocation from the whole aboveground plant to the roots and rhizomes. The translocation of sulfur is discussed in conjunction with the distribution and temporal dynamics of sulfur (see Chapter 4, Section 4.3.4). The second experiment, conducted in 1978, exposed individual leaf blades of *A. smithii* to $^{14}CO_2$ (Milchunas et al., 1981b). The labeling of individual leaves, by three leaf-age classes, enabled us to assess carbon translocation to specific aboveground organs as well as to belowground compartments; and to do so with reference to leaf age/position.

The translocation rate of ^{14}C to belowground plant components, when the total aboveground biomass was labeled, was highest in July irrespective of depth or treatment (Figure 5.10). Translocation to roots in the 0- to 5-cm layer was greater in May and July than in September, while below 5 cm the translocation to roots

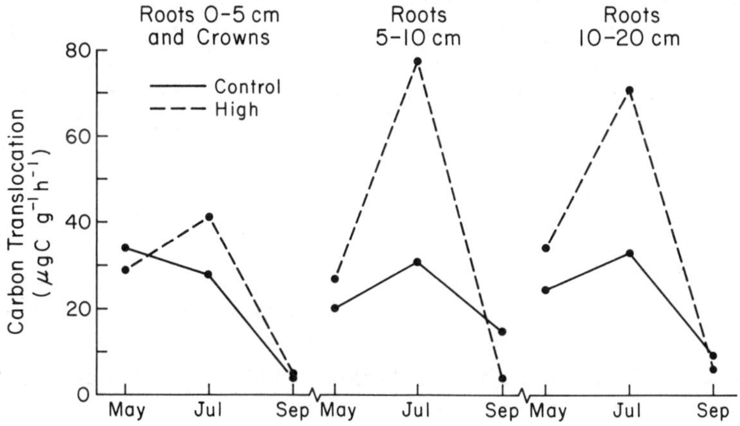

Figure 5.10. Belowground ^{14}C translocation rates ($\mu g\ C \cdot g^{-1} \cdot h^{-1}$) on Site I, 1976. (Adapted from Coughenour et al., 1979.)

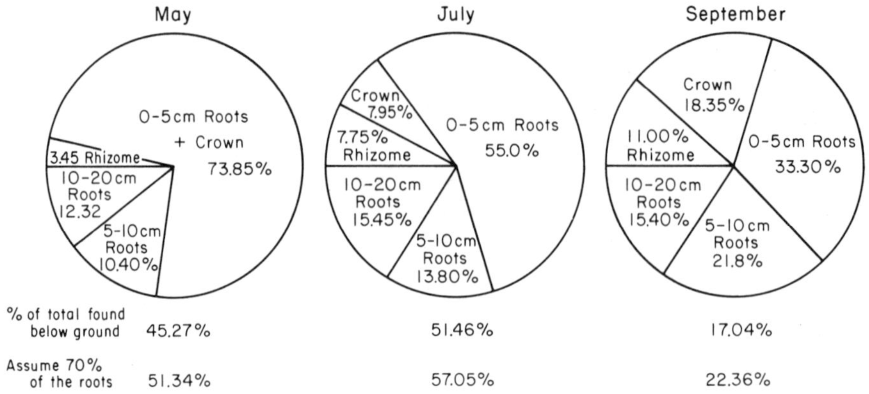

Figure 5.11. Partitioning of ^{14}C translocated belowground (%) in May, July, and September, Site I, 1976. The percentage of total [^{14}C]carbon found belowground is probably underestimated and might be adjusted on the assumption that 70% of the total roots were harvested at 0–20 cm. (Adapted from Coughenour et al., 1979.)

was greater in July. A SO_2 treatment effect was noted only in July. At this time aboveground biomass was lower on treated plots, yet translocation rates to roots were higher in both the 0- to 10-cm and 10- to 20-cm layers.

The percentage of total ^{14}C found belowground (Figure 5.11) declined from 45% in May and 51% in July to 17% in September. If these figures were adjusted on the assumption that 30% of the roots were actually found below 20 cm, the proportion of total growth occurring belowground would be 51% in May, 57% in July, and 22% in September. A greater percentage of total belowground ^{14}C was found below 5 cm later in the season. A greater percentage was also found in rhizomes as the season progressed, and between July and September crowns also received an increased percentage. These results agree with those of Singh and Coleman (1974), who found maximum root growth in the 0- to 10-cm soil depth from late May to July and maximum growth in the 10- to 20-cm soil depth from late July to September. The results, however, conflict with their finding that relative allocation to crowns decreased with time. The findings here correspond to those of Warembourg and Paul (1977), who found decreased percentages of carbon in 0- to 10-cm roots in late season and in 10- to 25-cm roots in midseason. The partitioning of carbon observed here represents the relative increments in biomass between roots and shoots or between depths rather than the allocation of total assimilate.

We cannot draw the conclusion that SO_2 increased translocation to belowground organs based only upon the increase in carbon translocation to roots on SO_2-treated plots in July at 5–20 cm. However, in our second translocation experiment we also observed increased belowground carbon translocation in plants on SO_2-treated plots (Milchunas et al., 1981a).

Three days after exposure to $^{14}CO_2$, labeled leaves contained a higher concentration (CPM · mg^{-1}) and quantity (CPM) of ^{14}C than other aboveground components, regardless of date or SO_2 treatment (Table 5.4). Exposure to SO_2

significantly reduced the concentration and total quantity of ^{14}C remaining in labeled leaves, indicating the presence of high demand in other components. Import of ^{14}C by mature nonlabeled leaves accounted for 0.5% of the total aboveground ^{14}C in control tillers and 1% in those exposed to SO_2. Other aboveground parts, excluding the developing leaf, accounted for 17 and 22% of the aboveground ^{14}C for the control and SO_2-treated tillers.

The concentration and total quantity of ^{14}C remaining in labeled leaves were significantly influenced by the age class of the labeled leaf. Concentrations of ^{14}C in the labeled youngest, middle, and older fully expanded leaves were 79, 83, and 92% of the total aboveground concentration. Total quantities of ^{14}C retained in the labeled leaves showed the same pattern. The three leaf-age classes, from young to old, retained 67, 77, and 84% of the total aboveground activity, respectively. The large quantity exported from younger leaves may be a function of their age or proximity to the developing leaf.

In many cases the developing leaf was a strong carbon sink (Table 5.4). The quantity of ^{14}C in developing leaves was 1.7 times greater in SO_2-treated tillers than in control tillers. Young leaves rely upon translocated carbon until they develop sufficient photosynthetic capacity to satisfy their needs. Swanson et al. (1976), Thrower (1962), and Fellows and Geiger (1974) reported that peak demand for carbon by developing leaves occurred when the leaf was approximately 25% of its final length. We investigated the possibility that the greater ^{14}C activity in developing leaves on the SO_2 treatment was a function of leaf size rather than a treatment effect and found no significant differences. The greater translocation to developing leaves in tillers on the SO_2 plots was, therefore, a response to SO_2 exposure.

The greater sink demand of developing leaves in the plants exposed to SO_2 could be because of more rapid cell division and growth rate or because the slower development of photosynthetic capacity in developing leaves exposed to SO_2 requires greater or longer carbon import from mature leaves before they are self-sufficient. Increased translocation from mature leaves that are fixing less carbon is possible because sink demand can override the availability of assimilate in regulating translocation (Ho, 1979).

Translocation of ^{14}C belowground was influenced by SO_2 exposure, labeling date, and leaf-age class (Table 5.5). Concentrations of ^{14}C in roots and rhizomes increased, compared with the control, in June and August, although only the June data were statistically significant. The largest part of the increased concentration was found in rhizomes.

Two points are apparent from the growth and carbon uptake responses in conjunction with ^{14}C translocation. First, increased ^{14}C translocation with SO_2 exposure observed in this study was a response to stimulated growth rate. Second, productivity and carbon uptake measurements may not be as sensitive as translocation to low-level SO_2 exposure. Noyes (1980) found translocation was inhibited more than photosynthesis at SO_2 concentrations of 220 and 2300 $\mu g \cdot m^{-3}$ and inhibited similarly at a SO_2 concentration of 6600 $\mu g \cdot m^{-3}$. Teh and Swanson (1979) reported photosynthesis and translocation were inhibited, respectively, by 74 and 48% at a SO_2 concentration of 6600 $\mu g \cdot m^{-3}$. The

Table 5.4. Carbon-14 Relative Concentration and Partitioning in *Agropyron smithii* on the Control and High SO_2 Treatment[1]

			^{14}C Mean Activity (%)					
			Labeled Old Leaf		Labeled Middle Leaf		Labeled Young Leaf	
Date	Treatment	Plant Part[2]	CPM/mg	CPM	CPM/mg	CPM	CPM/mg	CPM
June 18	Control	Leaf 3	97.0	95.4	0.2	0.2	0.1	<0.1
		Leaf 4	0.6	0.8	77.5	82.2	0.4	0.1
		Leaf 5	0.4	0.6	0.9	0.9	82.8	76.5
		Devl. Lf.	0.1	3.0	3.4	9.3	8.7	17.0
		Other	1.1	1.0	18.0	3.6	9.1	6.3
	SO_2	Leaf 3	89.0	74.9	0.4	0.2	0.2	0.1
		Leaf 4	1.1	1.4	84.5	78.5	0.3	0.2
		Leaf 5	0.8	1.0	1.2	1.1	64.1	43.1
		Devl. Lf.	6.8	20.8	10.1	18.1	13.4	46.8
		Other	2.3	2.0	3.9	2.8	22.1	9.8
June 25	Control	Leaf 3	97.6	95.0	0.7	0.5	0.3	0.3
		Leaf 4	0.2	0.2	91.6	78.5	0.9	0.9
		Leaf 5	0.5	0.5	1.2	0.8	82.8	61.9
		Devl. Lf.	1.2	3.9	5.4	19.4	8.5	32.4
		Other	0.5	0.5	1.2	0.8	7.5	4.6
	SO_2	Leaf 3	91.3	87.7	1.9	1.6	0.5	0.5
		Leaf 4	1.6	1.5	74.9	67.7	0.3	0.3
		Leaf 5	0.8	0.8	2.4	3.0	64.3	64.2
		Devl. Lf.	2.9	10.5	6.2	20.1	8.2	25.2
		Other	3.4	1.9	14.6	7.7	26.9	9.0

July 15	Control	Leaf 5	95.2	87.2	0.8	0.7	0.2	0.2
		Leaf 6	1.1	1.0	89.8	76.4	0.4	0.3
		Leaf 7	0.3	0.2	2.1	1.6	90.1	77.1
		Devl. Lf.	3.2	11.5	6.2	20.9	5.7	20.9
		Other	0.2	0.1	1.1	0.4	3.6	1.6
	SO_2	Leaf 5	91.1	86.5	0.3	0.4	0.6	0.6
		Leaf 6	2.8	2.9	74.5	82.6	1.5	1.4
		Leaf 7	0.9	0.7	1.8	1.7	88.8	84.3
		Devl. Lf.	2.8	9.4	2.6	9.7	3.2	10.8
		Other	2.4	0.6	20.8	5.7	5.9	2.8
August 2	Control	Leaf 5	94.0	78.5	0.5	0.5	0.1	0.1
		Leaf 6	0.6	0.5	91.9	81.7	0.7	0.7
		Leaf 7	0.3	0.2	1.1	0.7	90.7	70.2
		Devl. Lf.	4.7	20.8	6.2	17.1	7.6	28.9
		Other	0.5	0.1	0.3	<0.1	0.8	1.1
	SO_2	Leaf 5	84.3	65.3	0.7	0.5	0.4	0.5
		Leaf 6	2.8	2.2	80.7	64.7	0.9	0.8
		Leaf 7	0.9	0.5	2.5	1.6	73.0	57.8
		Devl. Lf.	11.0	31.8	14.9	32.9	10.0	36.3
		Other	0.9	0.2	1.2	0.3	15.7	4.6

[1]Values are based on means for plants harvested 3 days after $^{14}CO_2$ assimilation. Adapted from Milchunas et al. (1981a).
[2]Leaf counts are from bottom to top of plant. The highest number represents the youngest fully expanded leaf. "Devl. Lf." refers to the new developing leaf. Other refers to the stem and leaves other than the three top leaves and the developing leaf.

Table 5.5. Relative Concentration of ^{14}C in Belowground Components of Plants on the Control and High-SO_2 Treatment[1]

			^{14}C Mean Concentration (%)		
Date	Treatment	Plant Part	Labeled Old Leaf (CPM/mg)	Labeled Middle Leaf (CPM/mg)	Labeled Young Leaf (CPM/mg)
June 18	Control	Root	0.9	0.9	0.5
		Rhizome	0.6	5.2	0.7
	SO_2	Root	9.2	3.0	3.1
		Rhizome	17.5	11.9	1.5
June 26	Control	Root	1.4	3.4	2.4
		Rhizome	2.3	2.7	3.1
	SO_2	Root	4.1	4.2	2.4
		Rhizome	3.3	2.7	4.2
July 15	Control	Root	1.6	2.7	2.6
		Rhizome	1.7	3.5	4.0
	SO_2	Root	3.9	2.3	2.9
		Rhizome	3.2	2.1	1.1
August 2	Control	Root	1.1	1.0	1.0
		Rhizome	4.2	3.9	6.6
	SO_2	Root	3.9	4.3	2.6
		Rhizome	6.7	10.3	6.1

[1]Adapted from Milchunas et al. (1981a).

inhibition of translocation with 220 $\mu g \cdot m^{-3}$ SO_2 reported by Noyes contrasts with the increased translocation we observed with 200 $\mu g \cdot m^{-3}$ SO_2 (high treatment). This may be attributed to several major differences between experimental designs. Noyes used 11-day-old hydroponically growing bean plants having one remaining leaf at the time of labeling and monitored translocation after the leaves were exposed to SO_2 for 2 hr. In contrast, we used field growing plants with the influence of intact aboveground sinks and continual SO_2 exposure. Plant species, environmental conditions, SO_2 dose duration and concentration fluctuations (with possible plant adaptation), and sink demands may have contributed to the different responses observed in the two studies. Noyes (1980) suggested that SO_2 inhibited sieve-tube loading. If this occurs, the results from this study suggest that for low SO_2 dose, stimulated sink strength can override the mechanism involved in sieve-tube loading inhibition.

The previous translocation study showed stimulated translocation to roots in the 5- to 20-cm layer on the high-SO_2 treatment in July. Although neither study alone presents conclusive evidence of an increase in belowground translocation with SO_2 exposure, together they suggest a stimulation in translocation to belowground organs on the high-SO_2 treatment. The very different ^{14}C-labeling procedures between the two studies, i.e., source leaf versus entire aboveground plants, prevents specific comparisons. Leaves high on the culm export mainly to developing leaves, whereas lower leaves supply proportionately more translocate to belowground organs (Mor and Halevy, 1979; Cook and Evans, 1976). Therefore, translocation to belowground organs when any of the top three leaves were labeled would not be representative of overall carbon dynamics between above- and belowground components. Further, comparisons between studies with respect to date or root depth are impossible because in the latter study labeled leaf-age class remained constant across dates, yet there were additional leaves below the labeled leaves as the season progressed. Although comparison between the two studies must be limited, the results are complementary and indicate an increase in translocation to aboveground sinks in high-SO_2 treatment tillers concurrent with an increase in relative root and rhizome ^{14}C concentrations. This suggests a stimulation in root, reproductive, and leaf growth on the high-SO_2 treatment. However, increased growth rates may be accompanied by increased senescence rates.

5.5 Leaf Area Dynamics

The impact of SO_2 on leaf area is perhaps the most extensively investigated topic in air pollution research. Much of this work has focused upon assessment of leaf area damage in terms of visible symptoms of SO_2 exposure. A common method used to quantify foliar injury is to rank symptoms using an arbitrary scale or to estimate the percentage of leaf area, leaves, or plants injured. Our approach to assessing the impact of SO_2 exposure upon the dynamics of plant growth utilized leaf area, leaf number, and root/shoot measurements.

In this section we synthesize the results of four studies. Heitschmidt et al. (1978) determined the live and dead leaf areas of individual *A. smithii* tillers on Sites I and II during the 1975 and 1976 growing seasons using a destructive technique. Rice et al. (1979) determined the total number and the number of fully senescent leaves of permanently tagged individual *A. smithii* tillers during the 1976 and 1977 growing seasons. Lauenroth et al. (1981a) placed hydroponically growing *Bouteloua gracilis* plants on the control and high-SO_2 treatments and measured live/dead shoot and root weights. Milchunas et al. (1981b) determined the live and dead leaf area for all leaves of permanently tagged individual *A. smithii* plants during the 1977 growing season within each SO_2 treatment, and during the 1978 growing season on fertilized plots within each SO_2 treatment.

Heitschmidt et al. (1978) reported a significant difference between SO_2 treatments only in August, when total leaf area of tillers on the high-SO_2 treatment was significantly greater than total leaf area of tillers on the control (Figure 5.12). The validity of the significant difference was questionable because of non-corresponding responses on the low- and medium-SO_2 treatments. Analysis of the relation between the size of the four oldest leaves and their relative growth rate produced no significant treatment effects. The suggestion of a significant increase in leaf area on the SO_2 treatments prompted an analysis of the total number of leaves per plant. Control plants had significantly fewer leaves than SO_2-treated plants by the end of the growing season (Figure 5.13). These data suggested an increase in the leaf area of SO_2-treated plants resulting from an increase in the number of leaves per plant, rather than from an increase in leaf size.

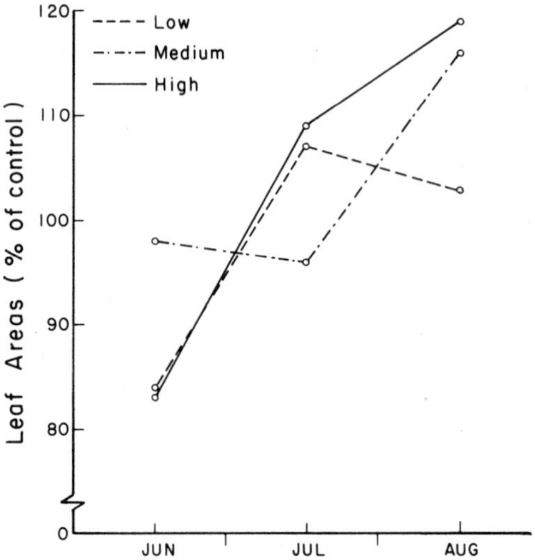

Figure 5.12. Effects of SO_2 treatments on leaf areas (% of control) of individual *Agropyron smithii* tillers for three dates in 1976, Site I. Vertical lines are LSR value at $P \leqq 0.05$. (Adapted from Heitschmidt et al., 1978.)

Previous studies have shown that exposure to SO_2 can hasten leaf senescence (Bleasdale, 1973; Matsushima and Harada, 1966). Dead leaf areas for *A. smithii* were 33, 28, 36, and 37% on the control, low-, medium-, and high-SO_2 treatments. None of these values were significantly different. However, significant SO_2 effects were present on some dates during the growing season. The percentage of dead leaf area of tillers on the control and low-SO_2 treatments in June was less than that on the medium- and high-SO_2 treatments. No significant differences were noted in August. Evidence that leaf senescence increased as a result of SO_2 exposure was

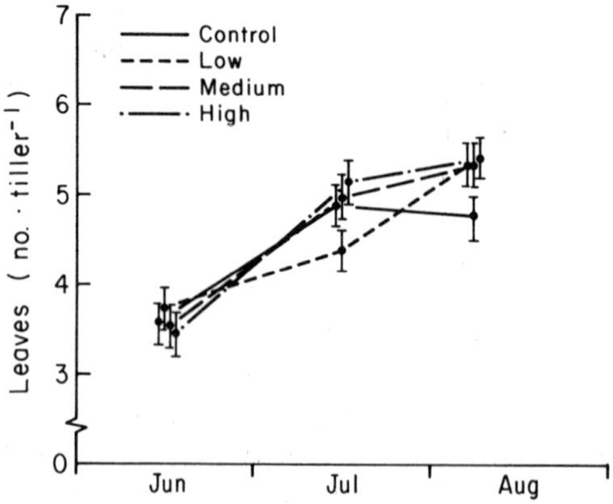

Figure 5.13. Effects of SO_2 treatments on the average number of leaves per *Agropyron smithii* tiller on each of three sample dates at Site I, 1976. Vertical lines are LSR value at $P \leqq 0.05$. (Adapted from Heitschmidt et al., 1978.)

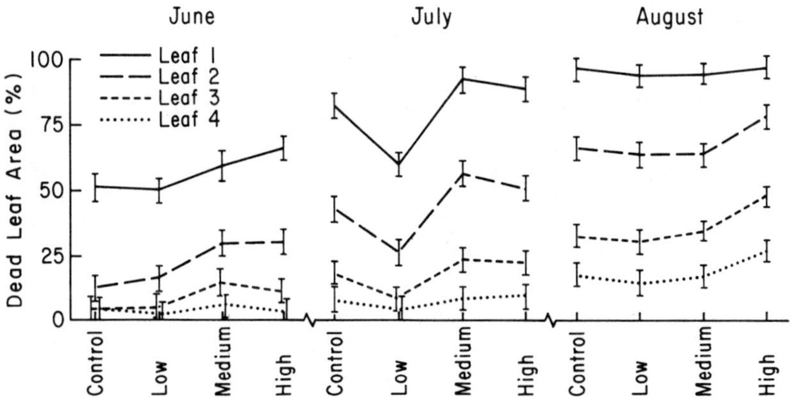

Figure 5.14. Effects of SO_2 treatments on the percentage of dead leaf area for the four oldest leaves on *Agropyron smithii* tillers, Site I, 1976. Leaf 1 is the oldest leaf. Vertical lines are LSR value at $P \leqq 0.05$. (Adapted from Heitschmidt et al., 1978.)

found only for the four oldest leaves of each plant (Figure 5.14). The proportion of dead leaf area of the four oldest leaves was 36, 31, 42, and 44% on the control and low-, medium-, and high-SO_2 treatments.

Counts of total leaves (Rice et al., 1979) revealed no significant SO_2 treatment effects on any of eight sampling dates. The ratio of dead to live leaves on Site I was significant only in September, when tillers on the high-SO_2 treatment had a higher proportion of dead leaves than tillers on the low-SO_2 treatment. On Site II, tillers on the high-SO_2 treatment showed significantly greater senescence than control tillers from early July through mid-August. The number of dead leaves on tillers in the high-SO_2 treatment were greater than on the low-SO_2 treatment from late June through early August, and greater than the medium-SO_2 treatment in early July. A trend was established with senescence greatest on the high-SO_2 treatment followed by the medium-, control, and low-SO_2 treatments. This trend continued until mid-August.

These studies focused upon the response of a C_3 grass to sulfur dioxide. Because stomatal behavior is important in determining plant responses to SO_2 exposure, the differences in stomatal control of photosynthesis observed between C_3 and C_4 plants (Körner et al., 1979) may result in quite different responses to SO_2 (Winner and Mooney, 1980; Carlson and Bazzaz, 1982). To test the hypothesis that a C_4 grass would respond differently from a C_3 grass we placed hydroponically growing *B. gracilis*, the most common C_4 grass in the northern mixed prairie, on each Site I treatment (Lauenroth et al., 1981a).

No significant treatment differences were observed in live *B. gracilis* shoot weights, root weights, shoot:root ratios, or number of tillers as a result of SO_2 exposure. Significant differences were found for the ratio of live to dead shoot weights. Plant situated on the low-SO_2 treatment had significantly greater live:dead ratios of shoot weights than on the medium- or high-SO_2 treatment (Figure 5.15). The pattern of senescence by treatment was high > medium >

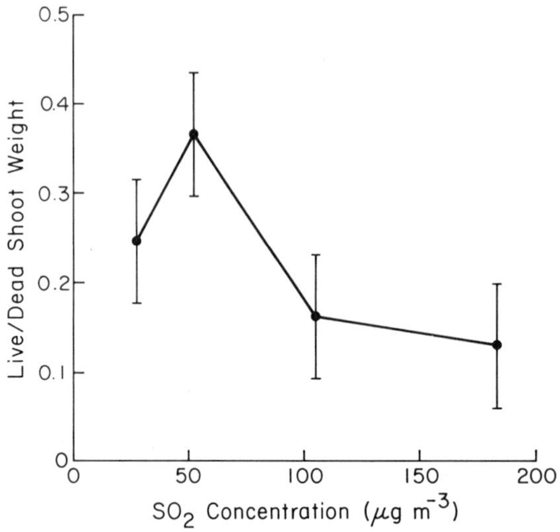

Figure 5.15. Ratio of live to dead shoot weights of hydroponically grown *Bouteloua gracilis* placed on the Site I SO_2 treatments, 1978. (Adapted from Lauenroth et al., 1981a.)

control > low SO_2. The hypothesis that differences in physiological behavior between C_4 and C_3 plants would result in different responses to SO_2 exposure was not supported by the results of this experiment. The responses of the C_4 grass *B. gracilis* were similar to previous responses for the C_3 grass *A. smithii*. This is in contrast to the results of Carlson and Bazzaz (1982), who found C_3 plants to be more sensitive than C_4 plants to SO_2 exposure at a concentration of approximately 650 $\mu g \cdot m^{-3}$.

The Heitschmidt et al. (1978), Rice et al. (1979), and Lauenroth et al. (1981a) studies suggested that both growth and senescence processes were altered by SO_2 exposure, but that the responses were subtle, and the low-SO_2 treatment could be stimulating growth. Because growth responses appeared subtle and contained both inhibitory and stimulatory elements, we initiated another experiment in 1978 which used a non-destructive sampling procedure whereby individual leaf blades of permanently marked tillers were measured throughout the growing season for live, dead, and litter (that portion of the leaf which because of its absence was assumed to have been added to the litter) area (Milchunas et al., 1981b). We also wished to assess the interaction of nitrogen and of sulfur fertilization with SO_2 (Milchunas et al., 1981b, Lauenroth et al., 1983). Fertilization treatments within the SO_2 treatments were control, nitrogen (150 kg nitrogen $\cdot$ ha^{-1} as ammonium nitrate), sulfur (15 kg sulfur $\cdot$ ha^{-1} as magnesium sulfate), and nitrogen plus sulfur (150 kg nitrogen plus 15 kg sulfur $\cdot$ ha^{-1}). Fertilizer was applied on 15 April.

Total leaf area for the SO_2 treatments without nitrogen fertilization did not differ significantly, although there was a trend toward greater leaf area with SO_2 treatment at peak and postpeak dates (Figure 5.16). With nitrogen fertilization, total leaf area on all SO_2 treatments increased and was doubled on the high-SO_2

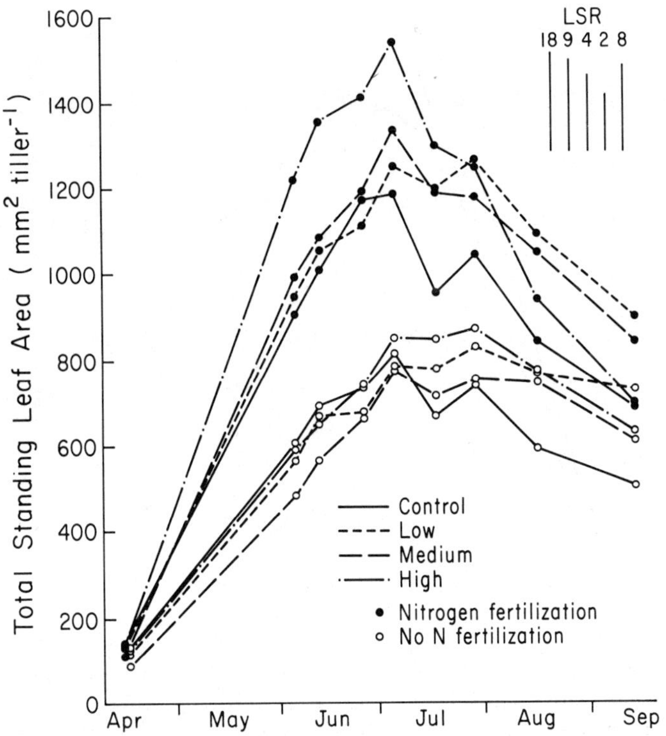

Figure 5.16. Total standing leaf area ($mm^2 \cdot tiller^{-1}$) of *A. smithii* on the SO_2 treatments through the 1978 growing season, Site I. Date $\times$ N $\times$ SO_2, $P \leqq 0.02$. Use LSR_2 for significance of N fertilization within SO_2 treatment within date, LSR_4 for across SO_2 within fertilization within date, LSR_8 for across SO_2 across fertilization within date, LSR_9 for within SO_2 within fertilization across date, and LSR_{18} for across SO_2 across or within fertilization across date. (Adapted from Milchunas et al., 1981b.)

treatment at the time of peak leaf area. Differences in leaf area after the peak decreased with time for SO_2 with nitrogen fertilizer compared with SO_2 without nitrogen fertilizer. Mineral nitrogen additions resulted in substantial differences among SO_2 treatments. Leaf areas on the high-SO_2 treatment were significantly greater than those of the control on all except the first and the last two dates.

The live leaf area of control plants was significantly lower than that of plants exposed to the high-SO_2 concentration, except on the first sample date when the plants were very small (Figure 5.17). The rate of increase before the maximum live leaf area was greater, and the decline after the peak was more pronounced, on the high-SO_2 treatment than on the low- or medium-SO_2 treatments. Nitrogen fetilization accelerated these increases and declines in live leaf area (Figure 5.18). Differences in live leaf area with and without nitrogen were insignificant by the September 15 sampling date.

Three-way interactions of date, nitrogen, and SO_2, though not significant at the tiller level, were significant when individual leaves were analyzed. Live leaf area of

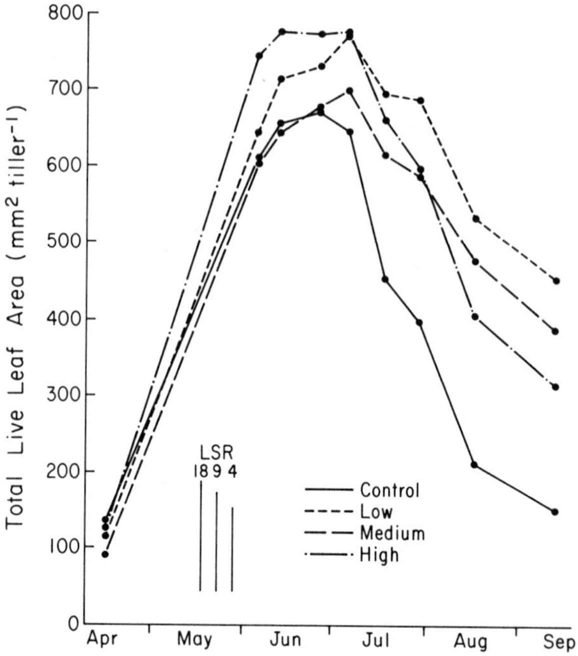

Figure 5.17. Total live leaf area ($mm^2 \cdot tiller^{-1}$) of *A. smithii* on the SO_2 treatments across all fertilization treatments, Site I, 1978. Date $\times$ SO_2, $P \leqq 0.001$. Use LSR_4 for significance of SO_2 treatments within date, LSR_9 for within SO_2 treatment across date, and LSR_{18} for across SO_2 treatment across date. (Adapted from Milchunas et al., 1981b.)

individual blades is a good indicator of the dynamics of growth because the value for any given treatment/date combination is not an average of all leaf-age classes, which include young, expanding leaves as well as old, senescing leaves. We examined data for leaf numbers 4 and 5 because their growth periods and our sampling dates best coincide to display the dynamics of growth. We will disuss two aspects of the interaction of date, nitrogen, and SO_2 treatment; the influence of nitrogen fertilization within SO_2 treatment, and the relationship among SO_2 treatments without N fertilization compared with the relationship among SO_2 treatments with N fertilization.

Within SO_2 treatments, peak live area of leaf 4 was much larger in nitrogen-fertilized plots than in non-N-fertilized plots (Figure 5.19). By 26 June, live area of leaf 4 was larger with N fertilization only on the low- and medium-SO_2 treatments. Near the end of the growing season, live leaf area on the low-SO_2 treatment was significantly greater without nitrogen fertilization. Nitrogen fertilizer induced a rapid decline in live leaf area on the high-SO_2 treatment that was not evident on the high-SO_2 treatment without nitrogen.

Sulfate fertilization had only a small influence on live leaf area (Figure 5.20). Significant interactions involving SO_4 were not observed for any other leaf area variable analyzed. Live leaf area of control tillers increased with sulfate-plus-nitrogen fertilization compared with nitrogen fertilization alone, but the increase

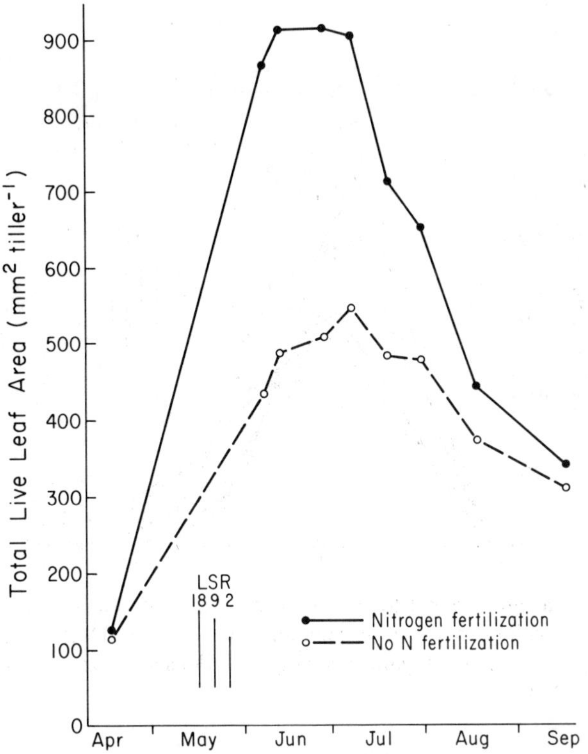

Figure 5.18. Total live leaf area ($mm^2 \cdot tiller^{-1}$) of *A. smithii* with and without nitrogen fertilization, across all SO_2 treatments, Site I, 1978. Date × N, $P \leqq 0.001$. Use LSR_{18} for significance of treatments across dates. LSR_9 for treatment within date, and LSR_2 for across treatment within date. (Adapted from Milchunas et al., 1981b.)

was not statistically significant. No interactions were observed between sulfate and SO_2.

The amount of leaf area transferred to the litter during the growing season was significantly influenced by date, nitrogen addition, and SO_2 exposure (Figure 5.21). Late-season litter production was significantly greater with N than without-N fertilization within SO_2 treatment and for the high-SO_2 treatment than for the other SO_2 treatments within N fertilization.

The cumulative number of leaves per tiller was not influenced by fertilization. The date by SO_2 treatment interaction, when the fertilization treatments within the SO_2 treatments were combined, suggested a greater average number of leaves per tiller with SO_2 at the end of the growing season (Figure 5.22). This trend was supported by the data of Rice et al. (1979) showing 6.7, 6.8, 7.3, and 7.1 leaves per tiller for the control and low-, medium-, and high-SO_2 treatments and by the Heitschmidt et al. (1978) data (Figure 5.13).

The number of leaves per tiller is a measure of subtle effects of SO_2 in our experiment. The response of leaf number to SO_2 treatments is unclear based on

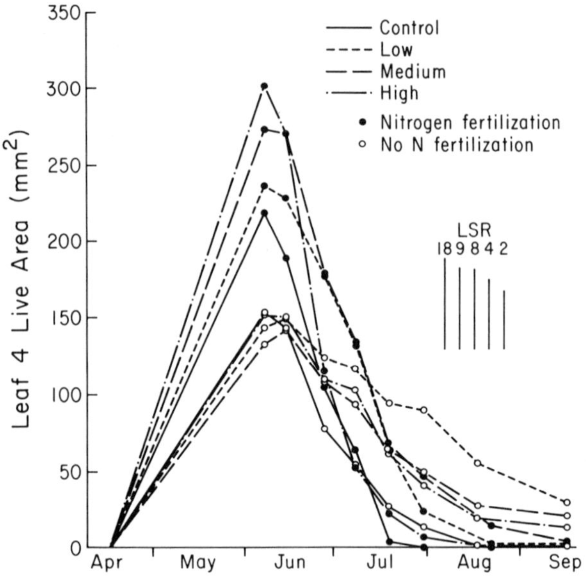

Figure 5.19. Live leaf area (mm^2) for leaf number four of *A. smithii* tillers on the SO_2 treatments with and without nitrogen fertilization, Site I, 1978. Date $\times$ N $\times$ SO_2, $P \leqq 0.001$. For explanation on use of significance ranges (LSR) see Figure 5.30. (Adapted from Milchunas et al., 1981b.)

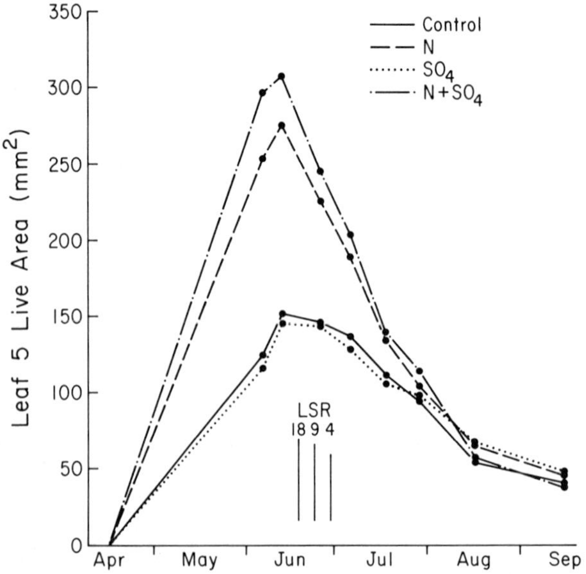

Figure 5.20. Live leaf area (mm^2) for leaf number five of *A. smithii* tillers fertilized with nitrogen and sulfur, across all SO_2 treatments, Site I, 1978. Date $\times$ N $\times$ SO_4, $P \leqq 0.036$. Use LSR_4 for significance of fertilization treatment within date, LSR_9 for across date within treatment, and LSR_{18} for across date across treatment. (Adapted from Milchunas et al., 1981b.)

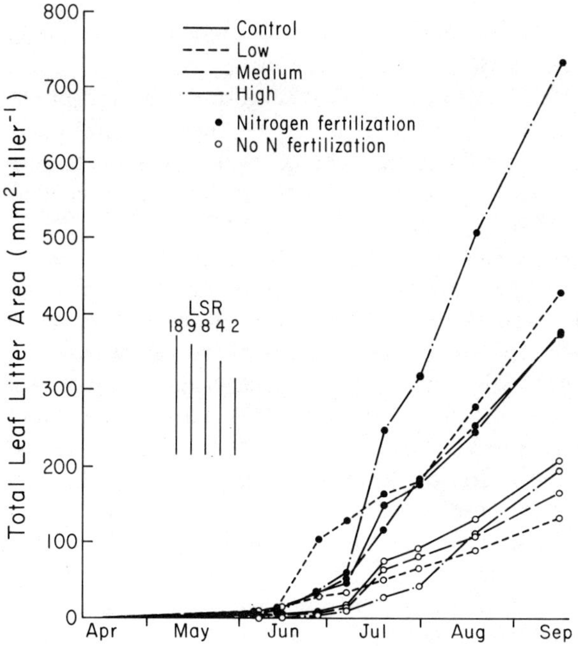

Figure 5.21. Litter production (mm^2 · tiller^{-1}) of *A. smithii* on the SO_2 treatments with and without nitrogen fertilization, Site I, 1978. Date × N × SO_2, $P \leqq 0.001$. For use of significance ranges (LSR) see Figure 5.17. (Adapted from Milchunas et al., 1981b.)

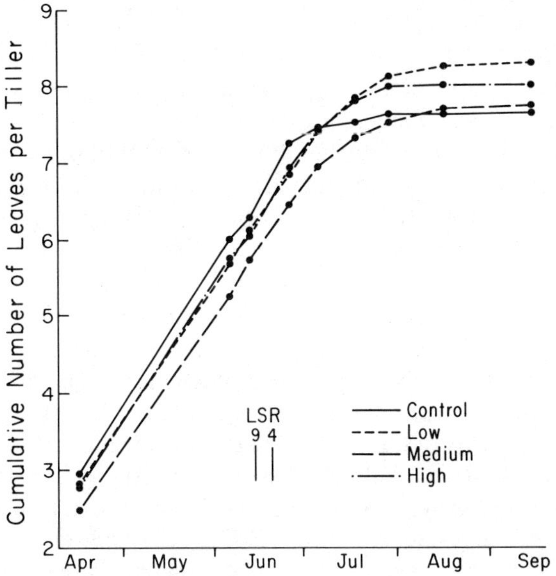

Figure 5.22. Cumulative number of leaves per *A. smithii* tiller on the SO_2 treatments, Site I, 1978. Date × SO_2, $P \leqq 0.001$. Use LSR_4 for significance of SO_2 treatments within date, and LSR_9 for dates within SO_2 treatment. (Adapted from Milchunas et al., 1981b.)

any one experiment, however, the three studies support the conclusion that SO_2 increased the number of leaves per tiller. An increase in the number of leaves per tiller with exposure to SO_2 was also observed by Ashenden and Mansfield (1977) and Bleasdale (1973). Bleasdale (1973) suggested that exposure to SO_2 enhanced cell division by interfering with the balance between oxidized and reduced sulfur radicals. This is not supported by reports of reduction in the number of leaves with SO_2 exposures of 288 $\mu g \cdot m^{-3}$ in *Dactylis glomerata* (Ashenden, 1978), and of 194 $\mu g \cdot m^{-3}$ in *Poa pratensis* (Ashenden, 1979).

Heitschmidt et al. (1978) concluded that the increased leaf area with SO_2 treatment resulted from an increase in the number of leaves. Milchunas et al. (1981b) also observed a concurrent increase in total leaf area and leaf numbers on the SO_2 treatments. Furthermore, the area of individual leaves did not increase with SO_2 treatment. The maximum leaf area of the two largest leaves averaged 179, 184, 131, and 134 mm^2 for the control and low-, medium-, and high-SO_2 treatments. The smaller size of individual leaves and the higher total leaf area per plant on the medium- and high-SO_2 treatments indicates that the increased size was a result of an increase in the number of leaves per tiller. This is substantiated by calculations of the average area per leaf showing 72, 92, 76, and 62 mm^2 for the control and low-, medium-, and high-SO_2 treatments. Both the number of leaves and the area of individual leaves increased on the low-SO_2 treatment. Tiller growth, as measured by live leaf area, significantly increased only on the low-SO_2 treatment. Exposure to low concentrations of SO_2 produced a subsidy effect, while the high-SO_2 concentration approached the level at which toxic effects began to appear. Further, the difference in the responses of total versus individual leaf area indicates a more rapid turnover of leaves on the high-SO_2 treatment.

The degree of senescence exhibited by *A. smithii* tillers on the SO_2 treatments correlates inversely with the leaf growth responses. The Heitschmidt et al. (1978), Rice et al. (1979) and Lauenroth et al. (1981a) studies ordered the extent of senescence by SO_2 treatment as high > medium > low. Milchunas et al. (1981b) observed less litter production and greater postpeak live leaf area on the low-SO_2 treatment than on the control, but similar responses between the control and high-SO_2 treatment. Again, the trends indicate a subsidy response with the low-SO_2 treatment, and toxic effects at the high-SO_2 concentration.

Although the expected growth increase with N fertilization was observed in this study, what is interesting are comparisons of growth patterns with and without N across date, SO_2 treatment, and SO_2 treatment and date. The maximum live leaf area of control tillers with N was significantly greater at the time of peak leaf area than live leaf area of tillers without N, but this did not continue throughout the growing season. Live leaf area of the SO_2 control N-fertilized tillers declined more rapidly than live leaf area of the SO_2 control non-N-fertilized tillers. Comparisons of SO_2 treatments with and without N show that total leaf areas converged but remained higher and litter production increased with N fertilization as the season progressed. Several important conclusions can be drawn from these results. First, ratios of live to dead leaf area or litter production suggest earlier senescence with N fertilization. Although rate of senescence is more rapid, total senescence is not greater. Rate of senescence may be a function of SO_2 subsidy rather than of toxicity. Toxicity can not be demonstrated unless absolute live leaf area is reduced. Second, these responses indicate that results based upon a single harvest at the end of the growing season will underestimate SO_2 effects when herbivory occurs throughout the season.

Growth responses were different on the SO_2 treatments with and without-N-fertilization. Nitrogen fertilization increased maximum leaf areas with increasing levels of SO_2. Total leaf area was greater on the high-SO_2 plus N treatment than the other treatments until the time of maximum leaf area, then declined more rapidly. Postpeak live leaf area on the high-SO_2 plus N treatment was also significantly less than on the low- or medium-SO_2 treatments. Leaf area lost to the litter was significantly increased by the high-SO_2 plus N treatment. These factors indicate that, with N fertilization, the rate of senescence was more rapid and time of senescence was earlier at the high-SO_2 concentration than at the low- or medium-SO_2 concentration. Because of the increased growth associated with SO_2 plus N fertilization, the observed senescence pattern may be the result of a shift in phenology with earlier development and senescence, as was observed with N fertilization alone, rather than a toxic effect, which would depress live leaf areas and growth. Distinguishing between rate and time of senescence is necessary when examining the effects of any compound that can be both a fertilizer and a toxic substance.

Cotrufo and Berry (1970) also observed higher resistance to SO_2 in fertilized plants. Application of 0.5 g of NPK fertilizer per pot sharply decreased pine needle injury from SO_2. With 1 g of fertilizer per pot, even less injury was observed. However, with 2 g of fertilizer per pot, tip necrosis was again evident. They could not explain the increased sensitivity to SO_2 with large fertilizer applications but suggested an interaction between the high salt concentration of the needles and air pollution. In our study, application of nitrogen was constant across three concentrations of SO_2. We also observed that nitrogen fertilization ameliorated the toxic effect of SO_2 but increased the rate and time of senescence on the high-SO_2 plus N treatment, indicating that the effect of the SO_2 by N interaction can be both positive and negative and that the threshold between the two can be altered by varying the concentration of either fertilizer or SO_2. Furthermore, whether the effects of any one combination of SO_2 and nitrogen are positive or negative depends on the duration of the treatments. The effects of high-SO_2 plus N were positive up to the time of maximum leaf area but negative afterwards.

The lack of a significant response to sulfate, without added N across SO_2 treatments is consistent with findings of Das and Runeckles (1975), Lockyer et al. (1976), Setterstrom et al. (1938), and Thomas et al. (1943). Faller (1971) reported minor yield variations in tobacco plants grown at 0, 40, and 80 ppm SO_4 and exposed to 1500 $\mu g \cdot m^{-3}$ SO_2 but a slight yield depression with a very high 240 ppm SO_4 treatment. Eaton et al. (1971) reported that growth was reduced by 37 and 54% in cotton and tomato, respectively, when SO_2 exposures at 200 $\mu g \cdot m^{-3}$ were supplemented with 300 M-equiv $\cdot$ $liter^{-1}$ SO_4. Growth response of tobacco exposed to 5240 $\mu g \cdot m^{-3}$ SO_2 was positive when SO_4 levels were increased from a suboptimal 1.5 ppm SO_4 to 96 ppm SO_4, but yield was depressed with 384 ppm SO_4 (Leone and Brennan, 1972). Our findings agree with those of

Lockyer et al. (1976), who found that concentrations of 0, 50, 100, 200, and 400 $\mu g \cdot m^{-3}$ SO_2 and 0 and 10 μg $SO_4 \cdot kg^{-1}$ dry soil produced no fertilizer treatment effect.

5.6 Biomass Dynamics and Net Primary Production

Continuous exposure to SO_2 during the growing season had no measurable impact upon plant phenology and minimal influences on the seasonal patterns of above- and belowground biomass or on estimated aboveground net primary production.

The basic patterns for live, recent dead, and old dead aboveground biomass observed for the control plots were indistinguishable from those observed for the SO_2 treatments (Figure 5.23). While slight differences in the timing of maxima and minima of the three biomass components are evident, none of these could be clearly identified as an impact of SO_2 exposure. Weather patterns and particularly the timing and amount of precipitation and its influence on soil water availability (see Figure 2.2) are the critical driving variables for aboveground biomass. Differences in either the timing or magnitude of biomass components could be explained by soil water.

The rates and amounts of biomass accumulation (live and recent dead) were similar for the three SO_2 treatments. Accumulation was rapid during April, May, and June, and leveled off or began to decline during July and August.

The single exception to the generalization of no effects of SO_2 treatments on plant biomass production was observed for a nonnative annual grass, *Bromus japonicus* (Figure 5.24) on Site II. This species was introduced in the northern Great Plains during the nineteenth century. *B. japonicus* most often germinates in the fall, overwinters in a vegetative stage, and completes its life cycle by early June. The most common control on the growth of *B. japonicus* over its life cycle is the presence or absence of soil water in the fall and early spring. Additionally, the potential of a particular site to support substantial populations of *B. japonicus* is believed to be related to the degree of disturbance of the indigenous vegetation. Biomass of *B. japonicus* on Site I was consistently low and showed no relationship with SO_2 treatments. Biomass was substantially greater on all plots on Site II presumably because of grazing history, and showed a relationship with SO_2 concentration in one out of three growing seasons. While similar responses of *B. japonicus* to SO_2 concentrations may have held for all growing seasons, the difficulty of sampling low biomass may have precluded detection.

A potential explanation for the sensitivity of *B. japonicus* compared with the other species in this grassland may be associated with SO_2 uptake rates. Gordon et al. (1978) reported sulfur accumulation rates for *B. japonicus* which were approximately double those measured for *A. smithii*. Chlorophyll content in *B. japonicus* was the most sensitive to SO_2 exposure of the eight species examined (see Figure 5.4).

Although biomass responses to the SO_2 treatments were not generally detected, responses of individual plant organs were observed (see Figures 5.16 and 5.17).

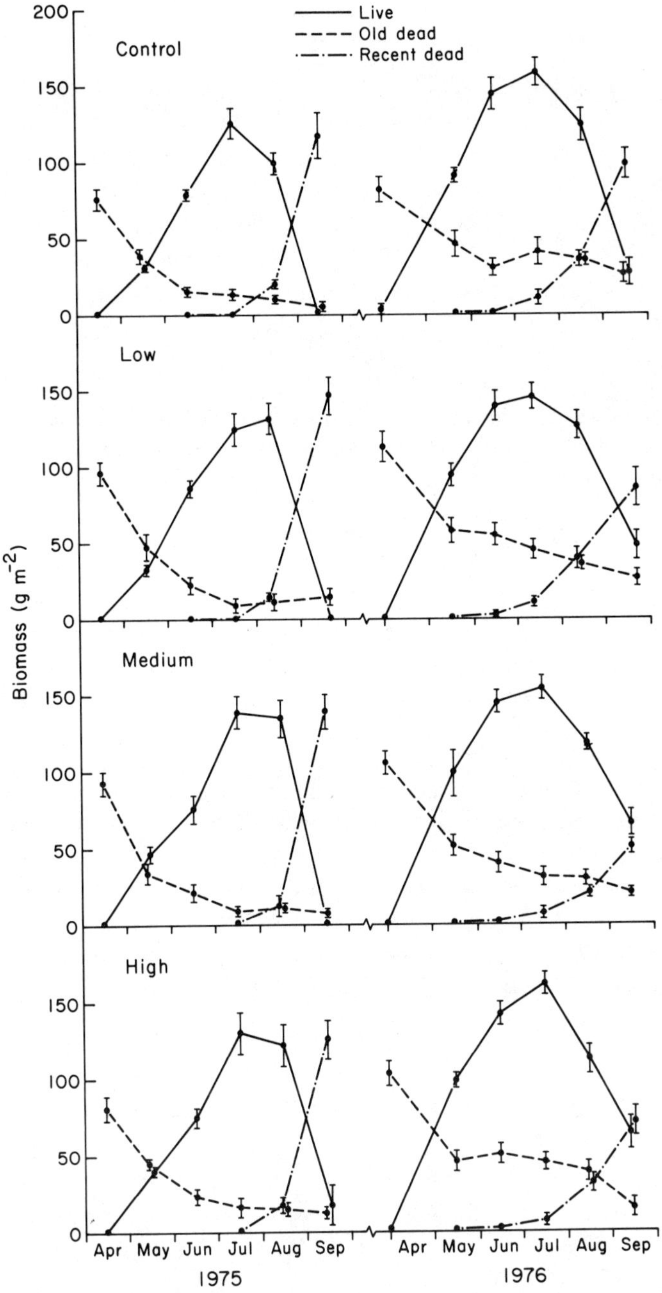

Figure 5.23. Live, recent dead, and old dead aboveground biomass (g · m^{-2}) on the control and three SO_2 treatments for 1975 and 1976, Site I.

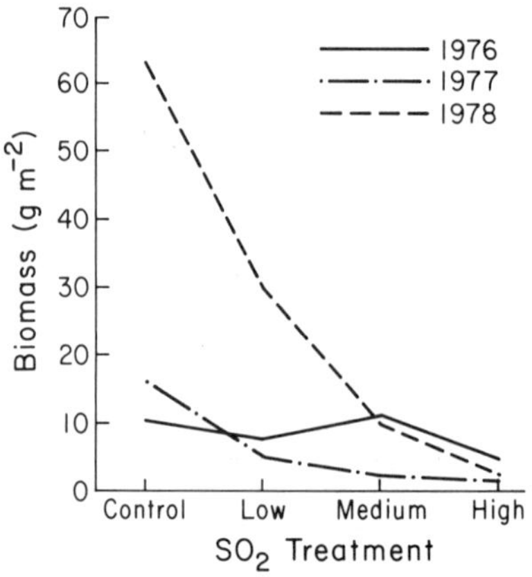

Figure 5.24. Biomass production (g · m^{-2}) for *Bromus japonicus* on control and three SO_2 treatments, 1976, 1977, and 1978, Site II. (Adapted from Dodd et al., 1982.)

Either technique sensitivity is greater at the organismal than at the population or community level, or the potential for adjustments in turnover of system components is large and effects at the lower levels are filtered out at higher levels.

Sulfur dioxide exposure had no measurable effect on aboveground net primary production (Table 5.6). Over the 4 years of measurement on Site I and 3 years on Site II, aboveground net primary production ranged from a low of 119 to a high of 268 g·m^{-2}·year^{-1}. Generally, Site I was less productive than Site II, presumably the result of greater fertility of the finer texture soils on Site II.

Belowground plant biomass trends through time are dominated by the growth of live roots and the decomposition of dead roots. Sulfur dioxide exposure had no measurable influence on root biomass (Figure 5.25). Root biomass was between 600 and 800 g·m^{-2} at the start and the end of the 1978 growing season. The

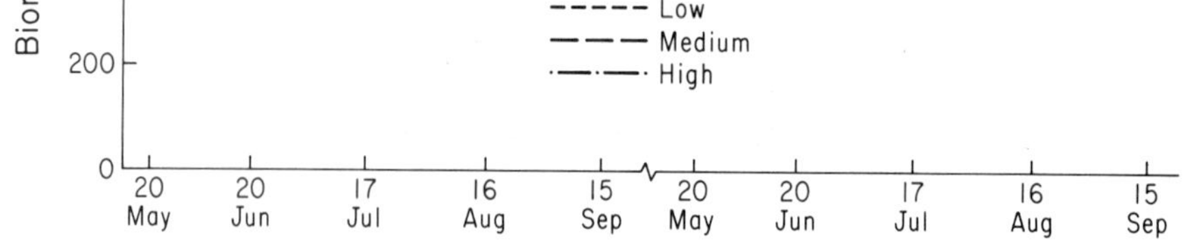

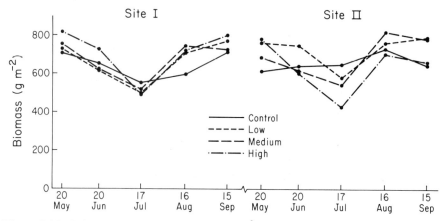

Figure 5.25. Belowground plant biomass (g · m^{-2}) to a depth of 10 cm through the 1978 growing season, Sites I and II.

Table 5.6. Aboveground Net Primary Production ($g \cdot m^{-2} \cdot yr^{-1}$) for Sites I and II, 1975–1978

	Site I				Site II			
Year	Control ($\bar{x} \pm$ SE)	Low ($\bar{x} \pm$ SE)	Medium ($\bar{x} \pm$ SE)	High ($\bar{x} \pm$ SE)	Control ($\bar{x} \pm$ SE)	Low ($\bar{x} \pm$ SE)	Medium ($\bar{x} \pm$ SE)	High ($\bar{x} \pm$ SE)
1975	150 ± 14	149 ± 17	165 ± 13	156 ± 14				
1976	199 ± 13	186 ± 9	195 ± 11	199 ± 8	177 ± 8	218 ± 17	205 ± 14	227 ± 15
1977	126 ± 12	131 ± 12	131 ± 7	119 ± 5	137 ± 7	210 ± 13	169 ± 9	190 ± 11
1978	136 ± 8	156 ± 14	123 ± 6	129 ± 6	268 ± 18	219 ± 18	214 ± 13	253 ± 8

Table 5.7. Production of Total and *Agropyron smithii* (Agsm) Biomass (g · m^{-2}) on Fertilized Subplots on Each SO_2 Treatment[1]

	Fertilizer Treatment							
	Control		Sulfur		Nitrogen + Sulfur		Nitrogen	
SO_2 Treatment	Total	Agsm	Total	Agsm	Total	Agsm	Total	Agsm
Control	172	74	118	28	240	66	368	126
Low	156	90	150	50	446	202	480	242
Medium	106	26	120	40	356	146	330	162
High	124	46	140	92	272	144	260	114

[1]The least significant range ($P \leqq 0.05$) for comparing total aboveground yield is 82 for SO_2 and is 76 for fertilizer. The least significant range for *Agropyron smithii* is 77 for comparing both SO_2 and fertilizer treatments. (Adapted from Lauenroth et al., 1983).

midseason decrease in biomass was presumably the result of decomposition occurring more rapidly than root growth. The pattern was independent of SO_2 exposure concentration.

Nitrogen and sulfur fertilizers as well as defoliation treatments were employed in an effort to further evaluate plant biomass and production as a result of SO_2 exposure (Lauenroth et al., 1981b, 1983). The objectives for each experiment included the idea that with a secondary stressor we could evaluate the state of the system with respect to thresholds which would clarify impacts of SO_2 exposure.

Nitrogen fertilization both with and without sulfur fertilization significantly increased aboveground biomass production regardless of SO_2 treatment (Table 5.7). Greatest production of all species was associated with nitrogen fertilization and season-long exposure to the low-SO_2 treatment. For both total and *A. smithii* aboveground biomass production, the nitrogen by SO_2 and nitrogen plus sulfur by SO_2 responses were parabolic across the SO_2 concentration gradient with maxima at the low-SO_2 treatment.

Belowground biomass was significantly affected by fertilizer treatments and the interaction of nitrogen additions with SO_2 (Table 5.8). Highest estimates were associated with treatments that received supplemental nitrogen and lowest with the treatment receiving supplemental sulfur. The responses to SO_2 plus nitrogen were similar to those observed aboveground. No significant differences were observed in belowground biomass on the control fertilizer plots, regardless of SO_2 concentrations.

Livestock grazing is the primary agricultural use of northern Great Plains grasslands. Several potentially conflicting results about the influence of SO_2 exposure on belowground plant organs led us to a hypothesis about the impact of SO_2 on the response of *Agropyron smithii* to defoliation. Measurements of the biomass of rhizomes at Site I indicated that SO_2 was inhibiting a recovery of rhizomes as a result of protection from grazing by cattle (see Table 5.3). Data from ^{14}C experiments suggested increased translocation of carbon compounds to belowground organs (Section 5.4). Since carbohydrates stored in belowground

Table 5.8. Belowground Biomass ($g \cdot m^{-2}$) for Fertilized Subplots on Each SO_2 Treatment[1]

SO_2 Treatment	Fertilizer Treatment: Control	Sulfur	Nitrogen plus Sulfur	Nitrogen
Control	956	940	1117	1301
Low	1135	932	1539	1207
Medium	863	730	1128	1100
High	1056	948	968	1203

[1]The least significant range ($P \leq 0.05$) for comparing SO_2 treatment means is 293 and for comparing fertilizer treatment means is 333. (Adapted from Lauenroth et al. 1983).

organs are required for growth initiation in the spring and after defoliation (Bokhari, 1977), an experiment employing defoliation treatments was chosen to clarify impacts of SO_2 on *Agropyron smithii.*

The potential for an SO_2–defoliation interaction was tested by defoliating subplots on the SO_2 treatments at two intensities and two frequencies (Lauenroth et al., 1982). Defoliation intensities were heavy (100% of aboveground biomass) and light (50% of aboveground biomass). The single defoliation occurred on 20 May, when aboveground live biomass of *A. smithii* is typically near 30% of the growing-season maximum. The second defoliation occurred near the time of expected peak live biomass (20 June). The plots were harvested on 15 August.

Agropyron smithii responded to SO_2 as a result of being defoliated. The single 20 May defoliation resulted in significant decreases in biomass at the medium- and high-SO_2 concentrations compared with the control (Figure 5.26b). Independent of SO_2, the single defoliation did not change the harvested biomass (Figure 5.26a).

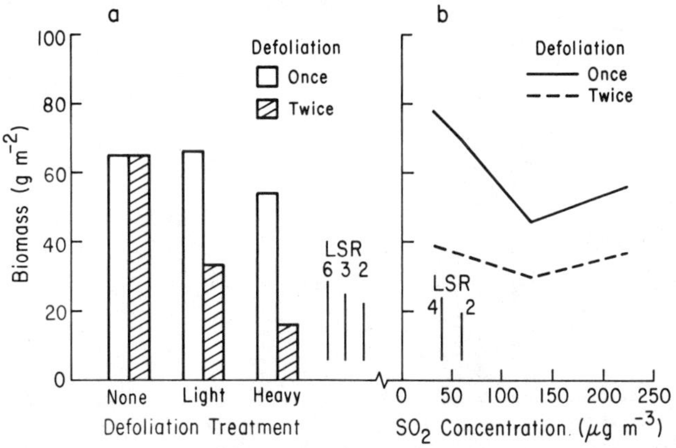

Figure 5.26. Response of *Agropyron smithii* biomass ($g \cdot m^{-2}$) (a) to three defoliation intensities and to two defoliation frequencies (b) to SO_2 concentration at two defoliation frequencies.

As a result of reapplying the defoliation treatments on 20 June, the differences among SO_2 treatments which were apparent after the single defoliation were no longer present, and independent of SO_2, the second light defoliation resulted in a decrease in biomass.

A change in biomass as a response to a treatment can occur because of changes in the size and/or number of individuals. Defoliation and its interaction with SO_2 also resulted in changes in *A. smithii* tiller density. Tiller density was decreased by all SO_2 treatments at both defoliation frequencies (Figure 5.27). Density responses to defoliation independent of SO_2 were different from those observed for biomass. Differences between defoliation frequencies at the light intensity, which had been observed for biomass, were not found for density. Tiller density increased with defoliation intensity and decreased only with the combination of two heavy defoliations.

In the previous section we reported no direct impact of SO_2 on either *A. smithii* biomass or density. The negative responses observed on SO_2 treatments that were also defoliated suggested that SO_2 impaired recovery from additional perturbation. Although the effects were subtle and not directly proportional to SO_2 concentration they indicated that regrowth potentials of grasslands exposed to SO_2 may be decreased as a result of grazing.

5.7 Summary

The responses of the mixed-prairie vegetation to exposure to SO_2 were variable in both time and space. Responses which were initially positive often became negative as time of exposure was extended. Responses that were significant at one site were not always substantiated by data from the other site. That we were able to continue the experiments for 5 years at two sites added substantially to the value of our results. Since we could evaluate specific responses over several growing

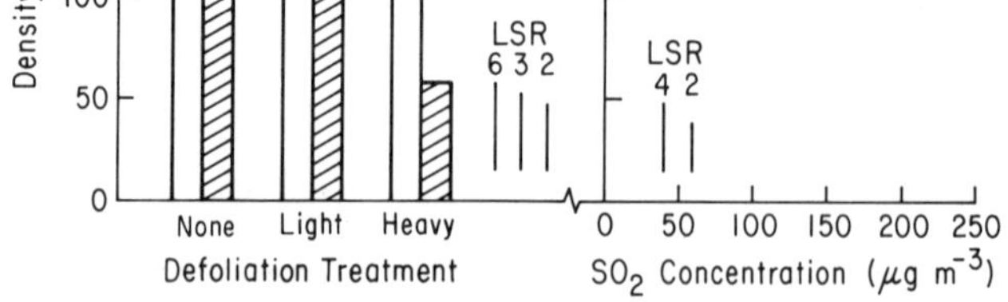

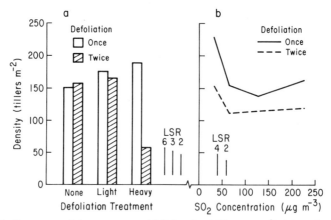

Figure 5.27. Response of *Agropyron smithii* density (tillers · m^{-2}) (a) to three defoliation intensities and to two defoliation frequencies (b) to SO_2 concentration at two defoliation frequencies.

seasons at each site we had the opportunity to consider ecological significance as well as statistical significance.

In addition to the temporal and spatial dimensions, we found that level of biological organization was important in influencing our results. The levels that we evaluated and examples of variables were: macromolecular (chlorophyll); cellular (plasmolysis in lichens); organ (leaf growth, rhizome biomass, etc.); individual (tiller development); population (tiller density); community (biomass, canopy cover, etc.); and system (net aboveground primary production). Statistically significant responses at lower levels of organization were often not verified at higher levels.

As a result of SO_2 exposure, we measured decreased chlorophyll concentrations in a variety of species including *Agropyron smithii*. The relationship between chlorophyll and carbon uptake is direct in terms of processes but not tightly coupled quantitatively. Impacts upon chlorophyll were not clearly translated to impacts upon plant biomass accumulation.

The translocation of carbon fixed in photosynthesis was influenced by SO_2 exposure. Labeling individual leaf blades with ^{14}C altered the distribution to other blades in a manner consistent with the impacts of SO_2 upon leaf growth. Changes in translocation to belowground organs were less clearly established but suggested an increase in translocation to roots and rhizomes at certain times during the growing season. Similar results were obtained when the entire aboveground portion of the plant community was labeled with ^{14}C.

Leaf growth and tiller structure for *A. smithii* were impacted by treatment with SO_2. The distribution of live and dead leaf material within the grassland canopy was altered as a result of SO_2 exposure. The turnover rate of leaves was increased as was the total number of leaves initiated on each tiller exposed to the highest SO_2 concentration. None of the changes in the structure of individual tillers influenced either their density per unit ground area or the contribution of *A. smithii* to canopy cover of the entire plant community.

Sulfur dioxide exposure had no measurable impact upon the amounts of above- and belowground plant biomass present at any time during the growing season, the rate at which biomass increased during a growing season, or aboveground net primary production. This generalization held for each of the years of measurement. The contribution of individual species and/or species groups to total biomass or production was altered by SO_2 exposure only in the case of the introduced annual grass *Bromus japonicus*. Biomass of *B. japonicus* is extremely variable from year to year depending upon weather patterns. The significance of this impact of SO_2 is unclear.

Secondary stressor experiments utilizing either fertilizers (nitrate and sulfate) or defoliation clarified several initially confusing responses but did not result in any fundamental changes in the observed impacts of SO_2. Conclusions from the fertilizer experiment were that nitrate additions can both ameliorate and induce negative effects of SO_2 exposure. The defoliation experiment indicated that while a small potential for a negative interaction with SO_2 is possible, *A. smithii* has a large capacity to adjust to both SO_2 exposure and defoliation.

The SO_2 concentrations chosen for this experiment were expected to produce a gradient of impacts on the grassland plant community ranging from minor to severe.

While a gradient was observed, the range of responses was quite small in most cases. The vegetation of the northern mixed prarie had a much greater capacity to adjust to SO_2 exposure than we had anticipated.

References

Ashenden, T. W. 1978. Growth reduction in cocksfoot (*Dactylis glomerata* L.) as a result of SO_2 pollution. *Environ. Pollut.* 15:161–166.

Ashenden, T. W. 1979. The effects of long-term exposure to SO_2 and NO_2 pollution on the growth of *Dactylis glomerata* L. and *Poa pratensis* L. *Environ. Pollut.* 15:161–166.

Ashenden, T. W., and T. A. Mansfield. 1977. Influence of wind speed on the sensitivity of ryegrass to SO_2. *J. Exp. Bot.* 28:729–735.

Bleasdale, J. K. A. 1973. Effects of coal-smoke pollution gases on the growth of ryegrass (*Lolium perenne* L.). *Environ. Pollut.* 5:275–285.

Buttery, B. R., and R. I. Buzzel. 1977. The relationship between chlorophyll content and rate of photosynthesis in soybeans. *Can. J. Plant Sci.* 57:1–5.

Carlson, R. W., and F. A. Bazzaz. 1982. Photosynthetic and growth response to fumigation with SO_2 at elevated CO_2 for C_3 and C_4 plants. *Oecologia (Berl.)* 54:50–54.

Coker, P. D. 1967. The effects of SO_2 pollution on bark epiphytes. *Trans. Br. Bryo. Soc.* 5:341.

Cook, M. G., and L. J. Evans. 1976. Effects of sink size, geometry, and distance from source on the distribution of assimilates in wheat. In *Transport and Transfer Processes in Plants*, I. E. Wardlaw and J. B. Passioura, eds. pp. 393–400. New York: Academic Press.

Cotrufo, C., and C. R. Berry. 1970. Some effects of a soluble NPK fertilizer on sensitivity of Eastern White pine to injury from SO_2 air pollution. *Forest Sci.* 16:72–73.

Coughenour, M. B., J. L. Dodd, D. C. Coleman, and W. K. Lauenroth. 1979. Partitioning of carbon and SO_2-sulfur in a native grassland. *Oecologia (Berl.)* 42:239–240.

Das, G., and V. C. Runeckles. 1975. Bisulphite-induced inactivation of growth and chlorophyll formation in *Chlorella pyrenoidosa. J. Exp. Bot.* 26:705–712.

Dodd, J. L., W. K. Lauenroth, and R. K. Heitschmidt. 1982. Effects of controlled SO_2 exposure on net primary production and plant biomass dynamics. *J. Range Manage.* 35:572–579.

Eaton, F. M., W. R. Olmstead, and O. C. Taylor. 1971. Salt injury to plants with special reference to cations versus anions and ion activities. *Plant Soil* 35:533–547.

Eversman, S. 1978. Effects of low-level SO_2 in *Usnea hirta* and *Parmelia chlorochroa. The Bryologist* 81:368–377.

Faller, N. 1971. Plant nutrient sulfur—SO_2 vs. SO_4. *Sulfur Inst. J.* 7:5–6.

Fellows, R. J., and D. R. Geiger. 1974. Structural and physiological changes in sugar beet leaves during sink to source conversion. *Plant Physiol.* 54:877–885.

Geiger, D. R. 1979. Control of partitioning and export of carbon in leaves of higher plants. *Bot. Gaz.* 140:241–248.

Hanksworth, D. L., and F. Rose. 1970. Qualitative scale for estimating sulfur dioxide air pollution in England and Wales using epiphytic lichens. *Nature (London)* 227:145–148.

Heck, W. W., S. V. Krupa, and S. N. Linzon. 1979. Methodology for the assessment of air pollution effects on vegetation. Minneapolis, Minnesota: APCA Speciality Conference Proceedings. April 19–21, 1978.

Heitschmidt. R. K., W. K. Lauenroth, and J. L. Dodd. 1978. Effects of controlled levels of sulfur dioxide on western wheatgrass in a southeastern Montant grassland. *J. Appl. Ecol.* 14:859–868.

Hesketh, J. D. 1963. Limitations to photosynthesis responsible for differences among species. *Crop Sci.* 3:493–496.

Ho, L. C. 1979. Regulation of assimilate translocation between leaves and fruits in the tomato. *Ann. Bot.* 43:437–448.

Knabe, W. 1976. Effects of sulfur dioxide on terrestrial vegetation. *Ambio* 5:213–218.

Körner, Ch., J. A. Scheel, and H. Baker. 1979. Maximum leaf diffusive conductance in air plants. *Photosynthetica* 13:45–82.

Lauenroth, W. K., J. K. Detling, C. J. Bicak, and J. L. Dodd. 1981a. Response of *Bouteloua gracilis* to controlled SO_2 exposure. In *The Bioenvironmental Impact of a Coal-Fired Power Plant*, E. M. Preston, D. W. O'Guinn, and R. A. Wilson, eds. pp. 59–65. Corvallis, Oregon: U. S. Environmental Protection Agency, Corvallis Environmental Research Laboratory.

Lauenroth, W. K., J. K. Detling, and J. L. Dodd. 1981b. Impact of SO_2 exposure on the response of *Agropyron smithii* to defoliation. In *The Bioenvironmental Impact of a Coal-Fired Power Plant*, E. M. Preston, D. W. O'Guinn, and R. A. Wilson, eds. pp. 49–58. Corvallis, Oregon: U. S. Environmental Protection Agency, Corvallis Environmental Research Laboratory.

Lauenroth, W. K., and J. Dodd. 1981a. Chlorophyll reduction in western wheatgrass (*Apropyron smithii* Rydb.) exposed to sulfur dioxide. *Water Air Soil Pollut.* 15:309–315.

Lauenroth, W. K., and J. L. Dodd. 1981b. The impact of sulfur dioxide on the chlorophyll content of grassland plants. In *The Bioenvironmental Impact of a Coal-Fired Power Plant*, E. M. Preston, D. W. O'Guinn, and R. A. Wilson, eds. pp. 66–73. U. S. Environmental Protection Agency, Corvallis Environmental Research Laboratory, Corvallis, Oregon.

Lauenroth, W. K., D. G. Milchunas, and J. L. Dodd. 1983. Response of a grassland to sulfur and nitrogen treatments under controlled SO_2 exposure. *Environ. Exp. Bot.* 23:339–346.

Leone, I. A., and E. Brennan. 1972. Sulfur nutrition as it contributes to the susceptibility of tobacco and tomato to SO_2 injury. *Atmos. Environ.* 6:259–266.

Linzon, S. N. 1978. Effects of airborne sulfur pollutants on plants. In *Suflur in the Environment: II. Ecological Impacts*, J. O. Nriagu, ed. pp. 109–162. New York: Wiley.

Lockyer, D. R., D. W. Cowling, and L. H. P. Jones. 1976. A system for exposing plants to atmospheres containing low concentrations of sulfur dioxide. *J. Exp. Bot.* 27:397–409.

Malhotra, S. S. 1977. Effects of aqueous sulfur dioxide on chlorophyll destruction in *Pinus contorta. New Phytol.* 78:101–109.

Matsushima, J., and M. Harada. 1966. Sulfur dioxide gas injury to fruit trees. V. Absorption of sulfur dioxide by citrus trees and its relation to leaf fall and mineral content of leaves. *J. Jap. Soc. Hort. Sci.* 35:40–44.

Milchunas, D. G., W. K. Lauenroth, and J. L. Dodd. 1981a. The effect of SO_2 on ^{14}C translocation in *Agropyron smithii* Rydb. *Environ. Exp. Bot.* 22:81–91.

Milchunas, D. G., W. K. Lauenroth, J. L. Dodd, and T. J. McNary. 1981b. Effects of SO_2 exposure with nitrogen and sulfur fertilization on the growth of *Agropyron smithii. J. Appl. Ecol.* 18:291–302.

Mor, Y., and A. H. Halevy. 1979. Translocation of ^{14}C-assimilates in roses. I. The effect of the age of the shoot and the location of the source leaf. *Physiol. Plant.* 45:177–182.

Noyes, R. D. 1980. The comparative effects of sulfur dioxide on photosynthesis and translocation in bean. *Physiol. Plant Pathol.* 16:73–79.

Odum, E. P., J. T. Finn, and E. H. Franz. 1979. Perturbation theory and the subsidy stress gradient. *BioScience* 29:349–352.

Peiser, G. D., and S. F. Yang. 1977. Chlorophyll destruction by the bisulfite-oxygen system. *Plant Physiol.* 60:277–281.

Rao, D. N., and R. Le Blanc. 1965. Effects of SO_2 on the lichen algae with special reference to chlorophyll. *Bryologist* 69:69–75.

Reutz, W. F. 1973. The seasonal pattern of CO_2 exchange of *Festuca rubra* L. in a montane meadow community in northern Germany. *Oecologia (Berl.)* 13:247–259.

Rice, P. M., L. H. Pye, R. Boldi, J. O'Loughlin, P. C. Tourangeau, and C. C. Gordon. 1979. The effects of "low level SO_2" exposure on sulfur accumulation and various plant life responses of some major grassland species on the ZAPS sites. In *Bioenvironmental Impact of a Coal-Fired Power Plant*, E. M. Preston and T. L. Gillett, eds. pp. 494–511. Fifth Interim Report, Corvallis, Oregon: U.S. Environmental Protection Agency, Corvallis Environmental Research Laboratory.

Setterstrom, C., P. W. Zimmerman, and W. Crocker. 1938. Effect of low concentrations of sulfur dioxide on yield of alfalfa and crucifarae. *Contrib. Boyce Thompson Inst. Plant Res.* 9:179–198.

Sims, P. L., J. S. Singh, and W. K. Lauenroth. 1978. The structure and function of ten western North American grasslands. I. Abiotic and vegetational characteristics. *J. Ecol.* 66:251–285.

Singh, J. S., and D. C. Coleman. 1974. Distribution of photoassimilated ^{14}carbon in the root system of a shortgrass prairie. *J. Ecol.* 62:359–365.

Singh, J. S., W. K. Lauenroth, and D. G. Milchunas. 1983. Geography of grassland ecosystems. *Progr. Phys. Geogr.* 7:46–80.

Skye, E. 1968. Lichens and air pollution. *Acta. Phytog. Snec.* 52:1–123.

Swanson, C. A., J. Hoddinott, and J. W. Sij. 1976. The effect of selected sink leaf parameters on translocation rates. In *Transport and Transfer Processes in Plants*, I. F. Wardlaw and J. B. Passioura, eds. pp. 347–356. New York: Academic Press.

Taylor, J. E., W. C. Leininger, and M. W. Hoard. 1980. Plant community structure on ZAPS. In *The Bioenvironmental Impact of a Coal-Fired Power Plant*, E. M. Preston and D. W. O'Guinn, eds. pp. 216–234. Fifth Interim Report. Corvallis, Oregon: Corvallis Environmental Research Laboratory, U.S. Environmental Protection Agency.

Teh, K. H., and C. A. Swanson. 1979. Comparative inhibition of photosynthesis and translocation by sulfur doxide in bush bean. *Plant Physiol. Suppl.* 63:187.

Thomas, M. D., R. H. Hendricks, T. R. Collier, and G. R. Hill. 1943. The utilization of sulfate and sulfur dioxide for the sulfur nutrition of alfalfa. *Plant Physiol.* 18:345–371.

Thrower, S. L. 1962. Translocation of labeled assimilates in soybeans. II. The pattern of translocation in intact and defoliated plants. *Aust. J. Biol Sci.* 15:629–649.

Warembourg, F. R., and E. A. Paul. 1977. Seasonal transfers of assimilated ^{14}C in grassland: Plant production and turnover, translocation and respiration. In *The Belowground Ecosystem—A Synthesis of Plant-Associated Processses*, J. K. Marshall, ed. pp. 133–140. Range Sci. Dept. Sci. Ser. No. 26. Fort Collins: Colorado State University.

Winner, W. E., and H. A. Mooney. 1980. Ecology of SO_2 resistance. III. Metabolic changes of C_3 and C_4 *Atriplex* species due to SO_2 fumigations. *Oecologia (Berl.)* 46:49–54.

Ziegler, I. 1975. The effect of SO_2 pollution on plant metabolism. *Residue Rev.* 56:79–105.

6. Responses of Heterotrophs

J. L. Leetham, W. K. Lauenroth, D. G. Milchunas, T. Kirchner, and T. P. Yorks

6.1 Introduction

Heterotrophs play a direct and indirect role in the structural development and functional processes of a system (O'Neill, 1976; Lee and Inman, 1975; Kitchell et al., 1979). As consumers they are a channel of energy and nutrient flow and thus directly influence the functioning of the system. They indirectly affect the functioning of the system in a role as rate regulators and substrate transformers and transporters. Through other various activities they may modify the physical environment which in turn indirectly affects the functioning of the system. The response of plants to consumers can have an influence on the structure of the system.

The effects of SO_2 on the mixed-prairie heterotrophs are difficult to study because of the structural and dynamic complexity of the invertebrate community. Interpretation of the entire data set has been hampered by sampling constraints and variability. With respect to the invertebrates, we have concentrated on elaborating effects on community structure. The effects of SO_2 on comunity structure are discussed for three groups: (1) tardigrades, rotifers, and nematodes; (2) macroarthropods; and (3) microarthropods. Tardigrades, rotifers, and nematodes are considered separately because, as members of the soil water fauna, they may be relatively more susceptible to SO_2-induced changes in soil pH. The effects of SO_2 on invertebrate community dynamics are discussed with respect to above- and belowground arthropods. This approach was taken because of the different characteristics of the two groups (mobile versus relatively immobile) and the different modes and patterns of SO_2–sulfur deposition and accumulation in aboveground versus belowground plant organs and soil (see Chapter 4). Grasshoppers (Orthoptera:Acrididae) are discussed separately because relatively high-resolution data were gathered and because of their potential effects on the native forage resource. Invertebrate community organization is discussed in terms of trophic groups and species diversity. In addition, we have drawn together evidence of potential indirect effects on economically important ruminants.

6.2 Invertebrate Community Structure

Two broad groups of the invertebrates censused showed significant changes associated with the suflur dioxide exposure. The first represents the active types, those that are generally good fliers and capable of migrating from the experimental plots. Changes noted in this group could result from direct toxicity, but active avoidance is also possible. The second group represents those invertebrates that are relatively immobile and therefore incapable of migrating from the plots. Population changes in this second group are suspected to be caused by direct toxicity of SO_2 or its derivatives to the organisms themselves or to their food resources. It must be remembered that in a real world exposure situation, this latter group may expand to include the former, for insect wings alone are an unlikely means of escape from pollution sources extending across a wide area.

6.2.1 Tardigrades, Rotifers, and Nematodes

The most dramatic population changes among the immobile types were found among the three recognized genera of tardigrades on the sites (Table 6.1) (Leetham et al., 1981a). Density estimates from all sample dates show consistent trends of substantial population reductions in the high-treatment plots and variously reduced populations in the intermediate treatments. These trends were confirmed statistically ($P \leqq 0.10$) in 1979, when soil core sizes were increased, resulting in reduced zero counts and overall lower sample variability. Treatment comparisons for the July 1979 data showed tardigrade populations on the medium and high treatments to differ significantly from those on the low and control plots. The high treatment differed significantly from the other three plots in September.

Total frequencies of occurrence were also sharply reduced for tardigrades. Individuals or groups were recorded in 90 to 100% of the control samples, but in only 15 to 50% of the high-treatment samples. Similarly, significant population declines under SO_2 exposure were noted for the non-stylet-bearing nematodes. On Site II, populations were reduced on all three treatments below that of the control ($P < 0.03$), while on Site I, only the high treatment was significantly ($P < 0.04$) reduced below the control (Table 6.2). Depth by treatment interactions were significant on Site II but not on Site I. However, an examination of both

Table 6.1. Density Estimates of Tardigrade Populations on Both Sites[1]

	SO_2 Treatments				
	Control	Low	Medium	High	Sig. Level
Site I					
14 July 77	3760 (1400)	1160 (400)	1600 (510)	170 (120)	N.S.
08 July 78	4100 (1780)	1550 (320)	1330 (720)	0 (0)	N.S.
16 Sept 78	13,240 (5370)	16,690 (7630)	7130 (2260)	910 (470)	N.S.
21 June 79	8390 (1960)	3590 (620)	1440 (400)	750 (390)	<0.001
15 Sept 79	5220 (1190)	3320 (630)	3040 (940)	420 (160)	<0.001
Site II					
14 July 77	3760 (1850)	1270 (820)	3980 (2410)	770 (500)	0.090
08 July 78	1820 (950)	2650 (2160)	940 (650)	170 (120)	0.090

[1]Numbers per square meter with standard errors in parentheses.

interactions indicated that the treatment effects in both cases took place in the 0- to 10-cm layer and that, at least for this length of treatment, the deeper soil layer was essentially unaffected by the SO_2 exposure.

Statistically significant ($P < 0.01$) density increases were noted for stylet-bearing nematodes on the medium-treatment plot on Site I in 1977 (Table 6.3). Treatment by depth interactions showed that this increase was restricted to the top 10 cm and not generally over the entire 20-cm soil profile. This population increase was not recorded on Site II in 1977, nor on either site in 1978. Similarly, the total nematode community did not show statistically significant variations with treatment (Figure 6.1).

Table 6.2. Densities of Non-Stylet-Bearing Nematodes on Both Sites in 1977[1]

SO_2 Treatment and Site	0- to 10-cm Soil Depth	10- to 20-cm Soil Depth
Site I		
Control	1200	150
Low	1100	190
Medium	1100	220
High	640	81
Site II		
Control	1300	280
Low	800	280
Medium	710	290
High	970	220

[1]Figures are thousands of individuals per square meter.

Table 6.3. Density of Stylet-Bearing Nematodes on Site I in 1977[1]

SO_2 Treatment	0- to 10-cm Soil Depth	10- to 20-cm Soil Depth
Control	720	500
Low	780	720
Medium	1500	690
High	830	770

[1]Figures are thousands of individuals per square meter.

Rotifier densities were greater on both sites in 1978 than in 1977, with Site I showing the greatest increase, but no statistically significant treatment differences occurred on either site in either year (Figure 6.2). However, rotifer densities tended to increase on the low treatment and then decrease with increased SO_2 concentrations on both sites during the latter year of exposure. Lower populations were found directly under to SO_2 delivery orifices during both years, on both sites.

These three general groups (the tardigrades, the nematodes, and the rotifers) are members of the soil water fauna, and thus intimately affected by changes in that water (Wallwork, 1970). The input of SO_2–sulfur to the soil and the decomposition of vegetation containing large amounts of sulfur can cause various chemical

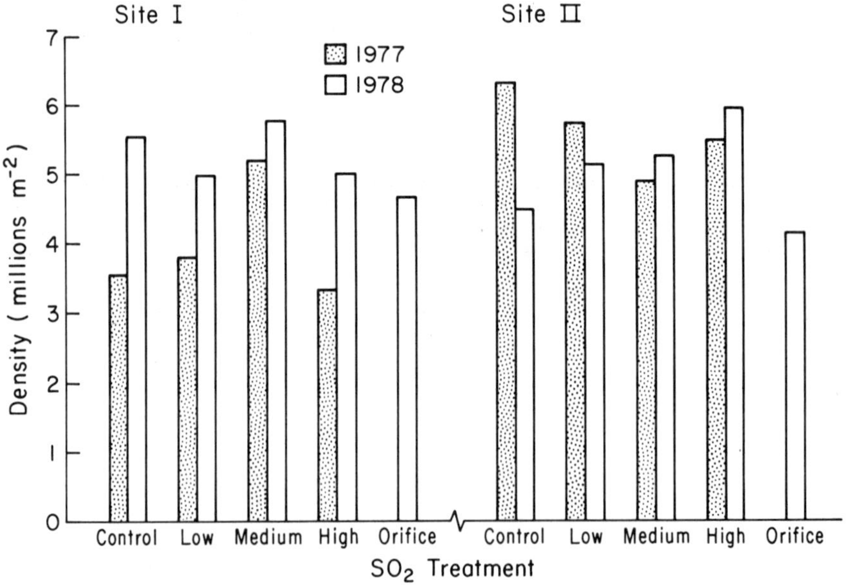

Figure 6.1. Nematode densities (all groups) (millions $\cdot$ m^{-2}) in the top 20 cm of soil on Sites I and II, 1977 and 1978.

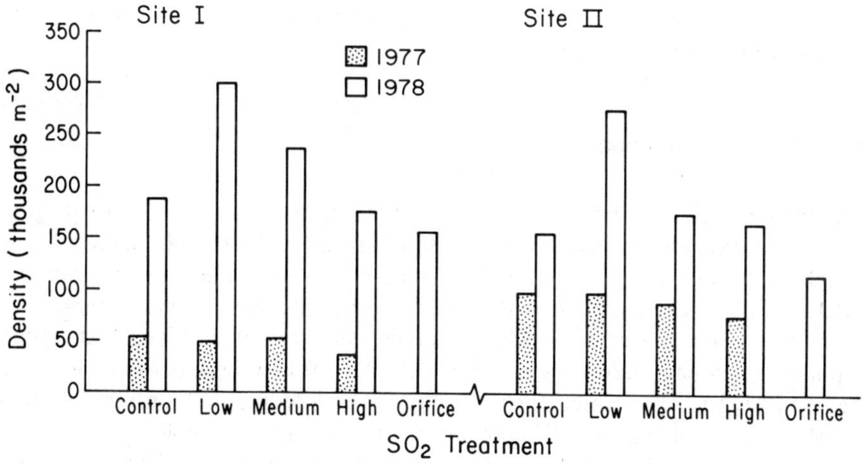

Figure 6.2. Rotifer densities (thousands · m^{-2}) in the top 20 cm of soil on Sites I and II, 1977 and 1978.

changes, including acidification. However, buffering may occur as the water percolates through the soil, and the chemical effects of the exposure may therefore be limited, in the short term, to the top 1 or 2 cm. Soil pH in the top 1 cm and litter disappearance rates were significantly reduced in the high-concentration SO_2 treatment (Dodd and Lauenroth, 1981; Leetham et al., 1983) (Table 4.11).

If this scenario is true, at least in this initial exposure phase, those organisms on or very near to the soil surface will be selectively affected by SO_2 exposure. Evidence which has so far been presented appears to support this concept, since the surface-dwelling tardigrades have shown the most severe effects, while the deeper ranging nematodes and rotifers have been less strongly affected. More difficult to evaluate are the long-term influences, for the soil buffering capacity is finite, and the current East Coast and Scandinavian acid precipitation controversies should raise cautions about any long-term conclusions. The prairie soils are often thought of as alkaline and heavily buffered, but, as Table 2.3 in Chapter 2 of this volume indicated, the critical top 25 cm of soil is slightly acidic.

6.2.2 Arthropods

The trends that were identified and confirmed statistically are outnumbered by trends that were not confirmed because of large sample variances. However, despite the numerous confirmed and apparent population changes as a result of SO_2 exposure, the major families were generally not among them, and the total arthropod density and biomass therefore appear unaffected during the experimental interval. Population changes did occur in a wide selection of groups, including the Acarina as well as across the spectrum from the primitive Collembola to the relatively advanced Diptera. These groups included representatives from both the immobile and mobile types. Nearly all the population trends, whether statistically significant or simply observable, showed decreases with increased SO_2 exposure concentration.

6.2.2.1 Macroarthropods

The beetles (Coleoptera) were the most consistently affected insect group. Significant population reductions were observed in the soil macroarthropod sampling for this group as well as for the aboveground sampling (Leetham et al., 1981c). Coleopteran families exhibiting declines included Chrysomelidae, Curculionidae, and Carabidae. The carabids, or ground beetles, had reduced populations on both field sites in 1975 and 1976. When these beetles appeared in the samples, they accounted for the bulk of the Coleopteran biomass, and hence the observed reduction in total biomass for the group as a whole.

The second most affected aboveground insect group was the Thysanoptera, or thrips. Significant population reductions were recorded for the high-treatment plots in 1976, while apparently similar, but not statistically significant reductions, occurred in 1975. These reductions were similar to those for the Coleoptera because in both cases the declines appeared to represent a cross section of the group rather than just one or two of the major species. Species identifications made on the samples were insufficient for statistical confirmation of any differential sensitivity.

While only a limited number of the aboveground arthropod groups showed statistically significant declines in population with exposure to SO_2, more showed simple trends in that direction. Since no similar list of groups showed an increase in population, the overall conclusion must be one of the potential for deleterious impact by SO_2 on aboveground arthropods.

6.2.2.2 Microarthropods

Many of the soil arthropod groups showed significant population reductions on the SO_2-treated plots, particularly in the second year of treatment on Site I (Leetham et al., 1981c). As with the aboveground arthropods, there were also additional apparent trends of population reduction that were not statistically significant. The observed trends of population reduction were noted in the early part of the growing seasons, at the time when soil moisture was at its highest. This agrees with the findings of Lebrun et al. (1978). Later in the season, with lower soil moisture and correspondingly low soil microarthropod populations and activity, SO_2 effects on this arthropod group appeared to be minor (Leetham et al., 1981b). Effects, therefore, may be activity related.

The most notable microarthropod reductions appeared in the Collembola during the second year on Site I. Significant reductions were noted in two of the four family groups identified, among the Poduridae and the Entomobryidae (Leetham et al., 1981b). A third group, the Sminthuridae, had a significant decline on Site II in the second year. The fourth group, the Onychiuridae, which are euedaphicly distributed deep in the soil profile, showed a nonsignificant decline.

It was pointed out in Chapter 2 (Section 2.2.4.1.2) that many of the soil microarthropods are distributed much deeper in the soil than the sampling reported here, which covered only the top 5 cm. These deep dwellers were not regularly sampled on the treatment plots, and any effects therefore remain unknown.

A counterpoint to the described population declines was found among the soil Acarina, where examples of significant population increases were found to be associated with SO_2 exposure. Five families showed such increases, though these held various inconsistencies among treatments and years. These results have been plotted for the fungivorous Gymnodamaeidae, the predatory Stigmaeidae, the fungivorous Tarsonemidae, and the predatory Phytoseiidae and Rhodacaridae (Figure 6.3).

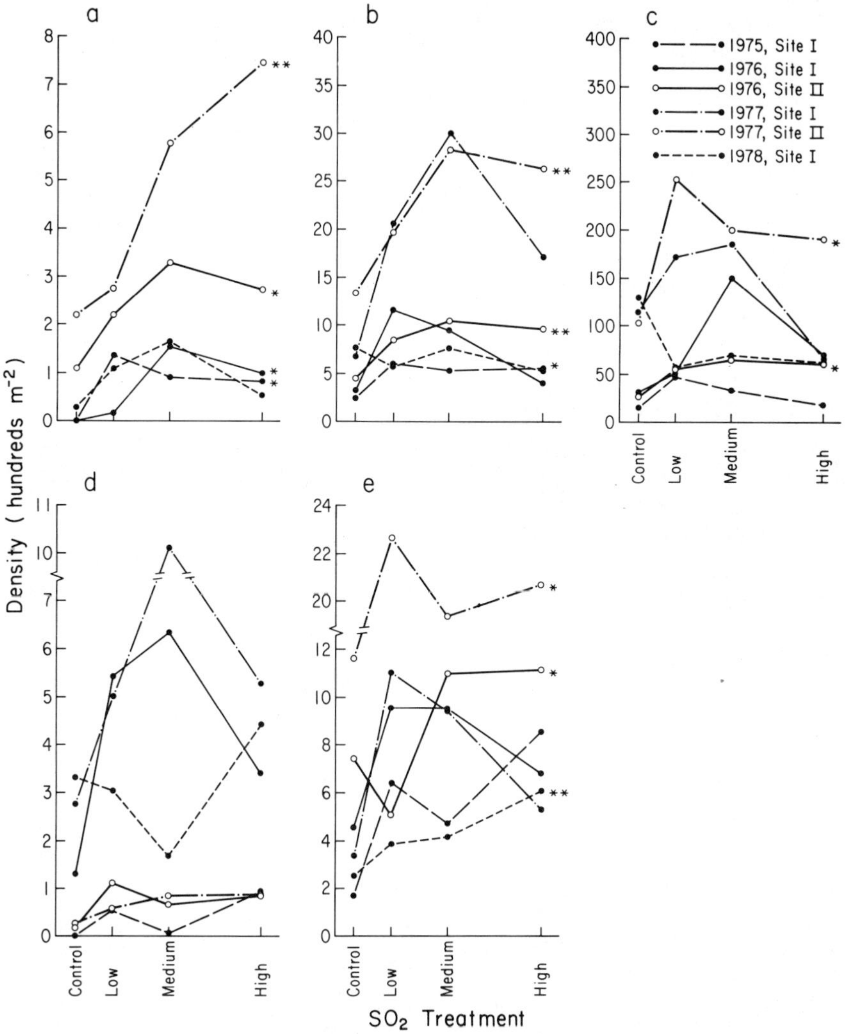

Figure 6.3. Densities (hundreds · m^{-2}) of five soil acarine families on Site I, 1975–1978; and Site II, 1976 and 1977. (a) Gymnodamaeidae; (b) Stigmaeidae; (c) Tarsonemidae; (d) Phytoseiidae; (e) Rhodacaridae. *Significant at $P \leq 0.05$, **Significant at $P \leq 0.01$.

The net result of these increases, together with the decreases already described, was no significant variation with treatment for either soil microarthropod total numbers or total biomass.

6.3 Invertebrate Community Dynamics

Quantifying the seasonal dynamics of invertebrates is extremely difficult for many reasons. Belowground, the major problems are high horizontal and vertical spatial variability as well as the high temporal variability caused by rapid turnover rates and inconsistencies between generations. The aboveground problems are the same but with important additional effect resulting from greater mobility.

6.3.1 Belowground Invertebrates

Several belowground microarthropod groups showed significant date by SO_2 treatment interactions. Collembola biomass was significantly reduced on all treated plots on Site I in 1976 (Figure 6.4). The largest portion of this response occurred in members of the family Entomobryidae. This was not, however, repeated in the other 3 years of data. In 1975, biomass was in the range of 4 to 7 $mg \cdot m^{-2}$ and was approximately equal among treatments. In 1977 and 1978, the biomass of the Collembola was less than 1 $mg \cdot m^{-2}$ and therefore too small for the detection of any treatment effects.

Three acarine families exhibited significant depressions in density during 1976 on Site I (Figure 6.5). Oppiidae and Oribatulidae had much higher numbers on the control plot than on any of the treated plots during the early part of the growing season. The differences decreased as the season progressed and the control total declined. The Cunaxidae showed significant differences only during the middle of

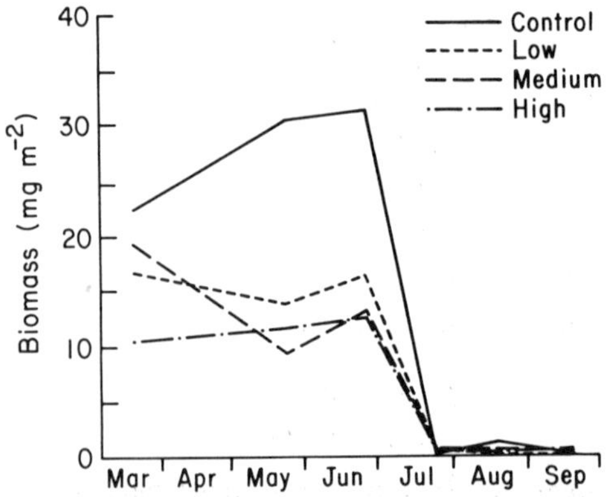

Figure 6.4. Seasonal dynamics of biomass ($mg \cdot m^{-2}$) of Collembola (total) on Site I in 1976.

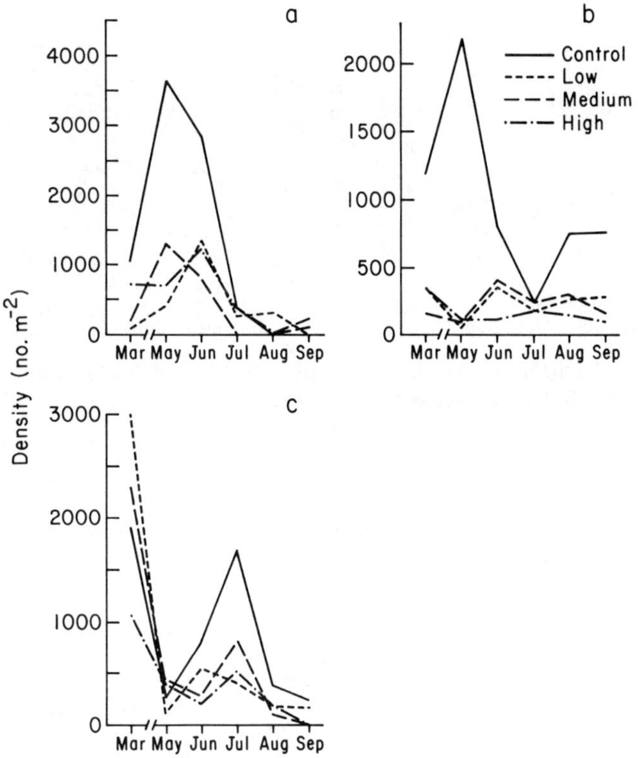

Figure 6.5. Density (No. · m^{-2}) of three acarine families on Site I in 1976. (a) Oppiidae. (b) Oribatulidae. (c) Cunaxidae.

the growing season. The significant results for this group were similar to the Collembola in being limited to one site in 1 year.

6.3.2 Aboveground Invertebrates

Aboveground arthropods were sampled only in 1975 and 1976. Significant date by SO_2 treatment interactions were found for Coleoptera density, and for the major Curculionidae family, on both sites and for both years, as illustrated for 1975 in Figure 6.6. Additional significant date by treatment interactions are presented in Figures 6.7 and 6.8. Conclusions about the effects of these alterations in invertebrate community dynamics are difficult to draw because the taxa of the two sites were different and, therefore, different taxa showed a response to SO_2.

6.3.3 Grasshoppers

Grasshoppers (with *Melanoplus* and *Eritettix* as example genera) can have a large impact on net primary production of grasslands (Mitchell and Pfadt, 1974).

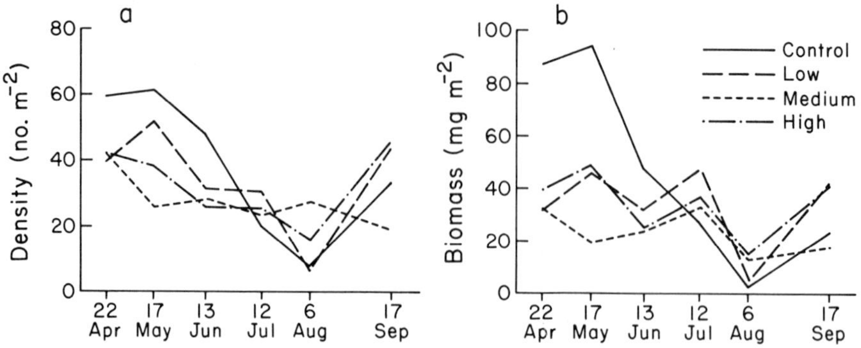

Figure 6.6. Seasonal dynamics of two insect groups on Site I in 1975. (a) Coleoptera (total) density (No. · m^{-2}). (b) Curculionidae biomass (mg · m^{-2}).

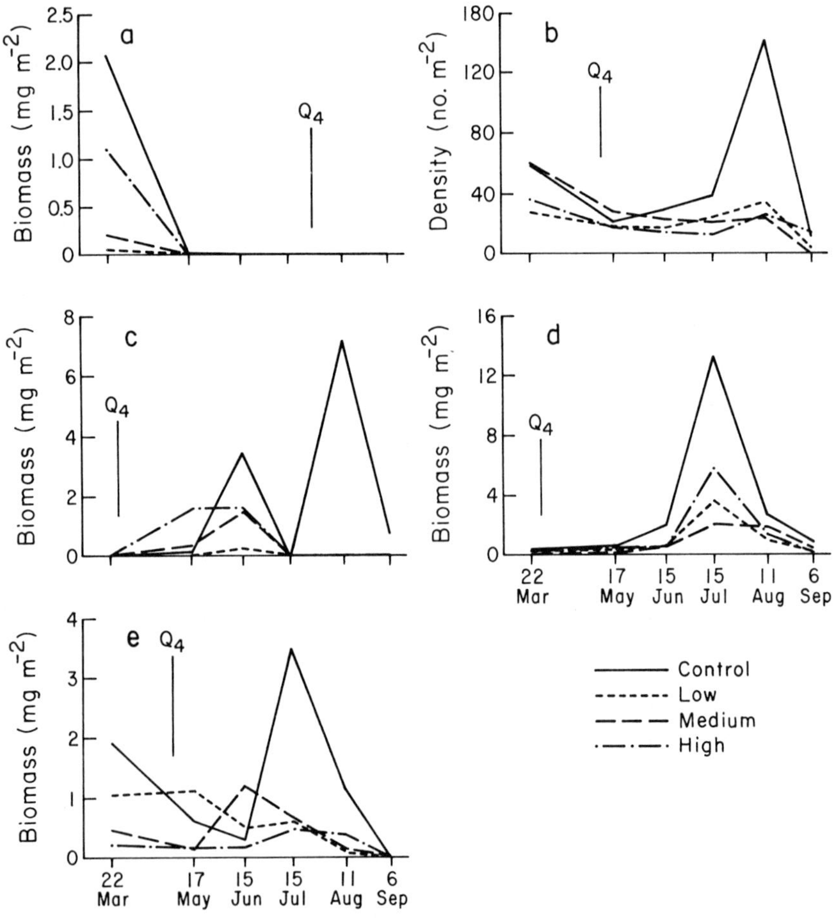

Figure 6.7. Biomass (mg · m^{-2}) or density (No. · m^{-2}) of five insect groups on Site I in 1976. (a) Poduridae (Collembola). (b) Homoptera (total). (c) Psyllidae (Homoptera). (d) Thripidae (Thysanoptera). (e) Diptera (total).

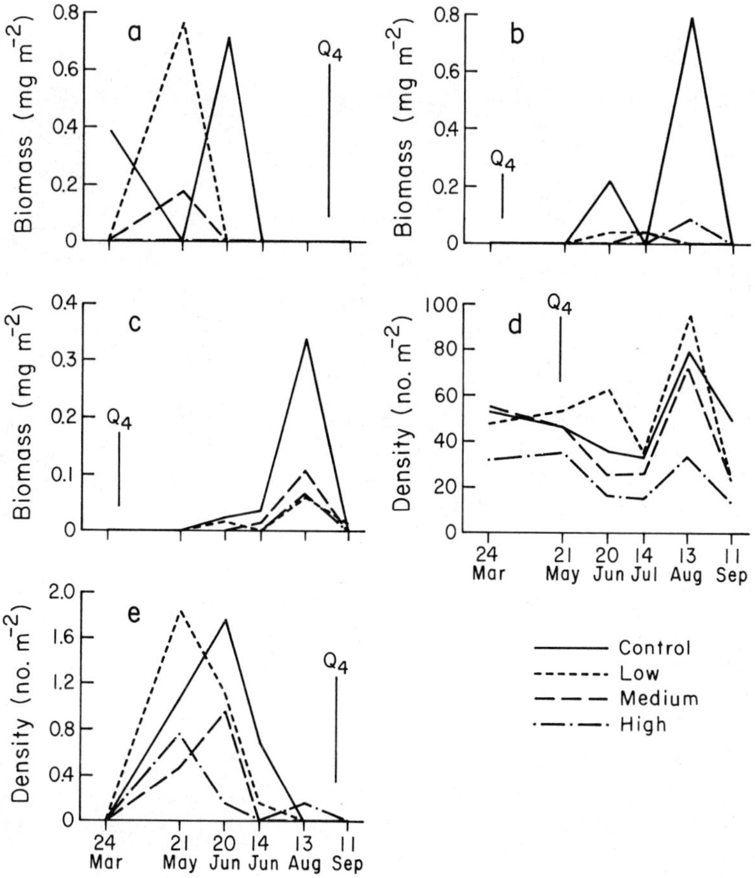

Figure 6.8. Biomass (mg · m^{-2}) or density (No. · m^{-2}) of five arthropod groups on Site II in 1976. (a) Geophilomorpha. (b) Araneida:Dictynidae. (c) Acarina:Oribatulidae. (d) Coleoptra (Total). (e) Hymenoptera:Diapriidae.

This is particularly relevant to the issue of SO_2 exposure because, as Mattson and Addy (1975) suggested, the activity of herbivorous insects appears to vary inversely with the vigor and productivity of the system, and insect outbreaks often occur in systems under stress. Sulfur dioxide emissions may be a system stressor. On the other hand, the direct toxicity of SO_2 to insects reported by Ginevan and Lane (1978) and Hillman and Benton (1972) together with the reduced palatability of forages with high-sulfur content may act to reduce grasshopper densities. Grasshopper responses to SO_2 were assessed using field flush censuses on the experimental sites and laboratory toxicity and diet selection tests.

The density of grasshoppers was significantly reduced ($P \leq 0.01$) by SO_2 treatments during late-growing-season dates within each year studied (McNary et al., 1981). The significant difference between the control and the high-treatment plot became apparent by 24 June in the first year and by 26 July in the second year (Figure 6.9). The decline in populations in the second year was likely a result of

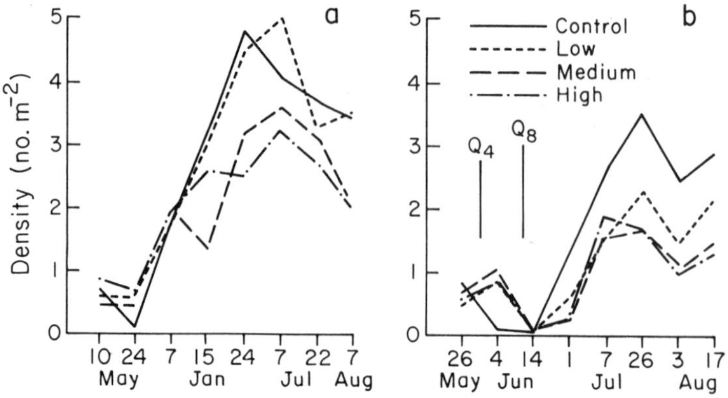

Figure 6.9. Density (No. · m^{-2}) of grasshoppers on Site I: (a) first year of sampling, third year of SO_2 exposure; (b) second year of sampling, fourth year of SO_2 exposure. Use Q_4 to test for significant differences between treatments within a date and Q_8 for within treatment across date comparisons. (Adapted from McNary et al., 1981.)

the wet and cold spring during that year. Grasshopper census data combined across replicates, years, and dates gave densities of 2.9, 2.4, 1.9, and 1.8 · m^{-2} for the control, low-, medium-, and high-treatment plots.

Grasshoppers classified as maturing late in the growing season composed 89% of the total observations. *Melanoplus sanguinipes* (F.) and *Eritettix simplex* (Scudder) were the two most common species, comprising 55 and 9% of the total. The within-year treatment by date response of *M. sanguinipes* was significant ($P \leq 0.01$) and displayed a pattern identical to the total grasshopper density, indicating the relative proportions of this species within the overall population were unaffected by SO_2.

Two possibilities were advanced for the observed grasshopper reductions on the experimental plots. There may have been a direct effect of the SO_2 on grasshopper reproduction or survival, or there may have been the indirect result of an increase in the emigration/immigration ratio for this highly mobile population. In laboratory experimentation, no significant differences were detected between control and SO_2 treatment at 470 μg · m^{-3} for grasshopper egg hatching success, mean development time for each nymphal instar, adult dry weight biomass, nor egg production per female per day (Leetham et al., 1980). However, there was a significant reduction in the variance of the developmental rate for the third, fourth, and fifth instars for the SO_2-exposed nymphs, a reduction that appeared to be associated with a significant increase in mortality for the exposed nymphs (Figure 6.10). Consequently, it was postulated that physiologically marginal individuals were more sensitive to SO_2 and were eliminated by the exposure.

That the second possibility was operational as well is suggested by the detection of fewer grasshopper eggs on the SO_2-treated plots. Density of eggs on the control plot at Site I were approximately 25 · m^{-2} compared with less than 10 · m^{-2} for the SO_2 treatments. A species of beetle, *Canthon* spp., was observed to avoid migrating onto the SO_2-treated plots (Bromenshenk and Gordon, 1978). This

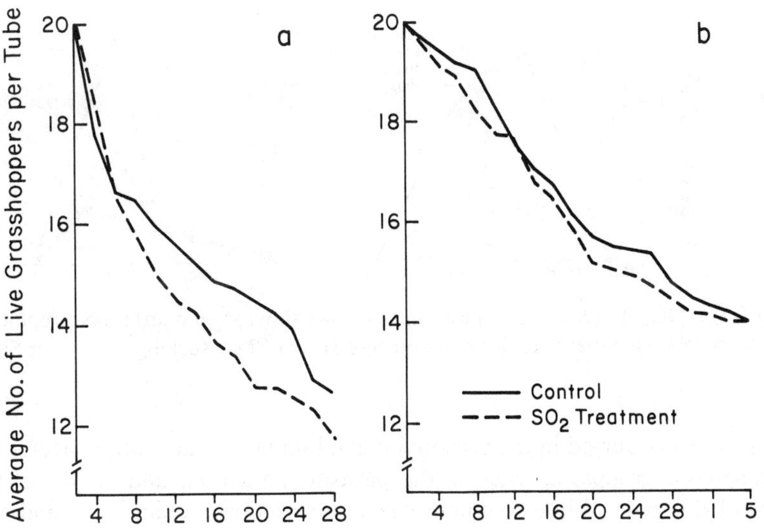

Figure 6.10. Grasshopper survival (No. live · tube^{-1}) through five nymphal instars to adult under control and SO_2 exposure (470 $\mu g \cdot m^{-3}$) in the laboratory. Trial 1 (a), and trial 2 (b).

avoidance response to SO_2 may be partially explained by reduced palatability of forages on the exposed areas. Cates and Orians' (1975) index of palatability was calculated using data from "cafeteria" feeding trials with SO_2-treated and untreated leaves of *A. smithii*. We found that male grasshoppers significantly preferred control leaves and a near significant preference for control leaves by the larger and less selective females over leaves exposed to sulfur dioxide.

6.4 Invertebrate Community Organization

6.4.1 Trophic Groups

Comparing trophic groupings, which combine taxa that have similar functions, suggested that SO_2 exposure may have reduced parasite–parasitoid (Figure 6.11a) and plant sucking insect (Figure 6.11b) populations and affected the timing of the former group. Parasite numbers were significantly reduced by exposure to the medium and high SO_2 concentrations. The exposures appeared to shift the population peaks to successively later in the season as the concentration increased. Both the changes in the timing of peaks and the reductions of absolute number may have consequent effects on the host populations. Plant sap sucking insects were reduced in number by all SO_2 exposures, but the timing of population peaks was unaffected, and the population reduction with exposure level was not as consistent across treatments as for the parasite group.

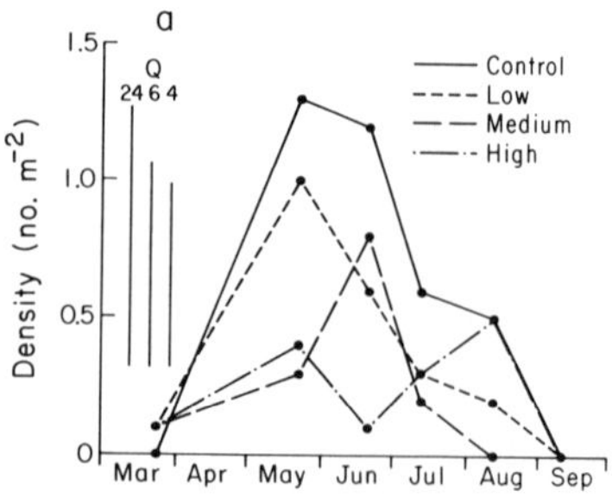

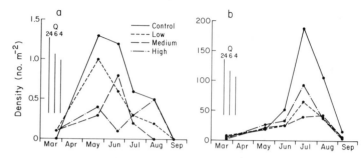

Figure 6.11. Density (No. · m^{-2}) estimates of two aboveground arthropod trophic groups in 1976. (a) Parasite–parasitoid insects on Site II. (b) Plant sucking insects on Site I.

Reductions occurred in the seasonal mean biomass of the pollen–nectar feeding and fungivore groups, as well as the parasite–parasitoid and plant sap feeding groups (Table 6.4). While community level interactions are intricate, and many of the pieces could not be accounted for by our observations, results for the first 2 years of exposure indicated that the arthropod community was responding in a mixed negative and positive fashion.

6.4.2 Diversity

Indices representing community diversity (Peet, 1974), information content (Pielou, 1966), equitability (Lloyd and Ghelardi, 1964), and richness (Hurlbert, 1971) were analyzed for each of the four treatments on both experimental sites using two-way, date by treatment, least-square analysis of variance for aboveground arthropods and soil microarthropods. Date of measurement had a significant effect ($P \leq 0.05$) for each of the diversity indices except richness of the aboveground arthropods on Site II. A reduction in information content based upon density on all four treatments on Site I coincided with a July outbreak of thrips. The outbreak extended across all four treatments on that site.

Table 6.4. Season Mean Population Estimates of Three Trophic Groups of Aboveground Arthropods and One Trophic Group of Soil Microarthropods on the Two Field Sites in 1976[1]

Group	SO_2 Treatment			
	Control	Low	Medium	High
Parasite–parasitoid biomass on Site II	1.7	1.0	0.3	1.0
Plant sucking insect biomass on Site I	110	75	65	83
Pollen-nectar feeding insect densities on Site II	2.7	4.7	2.6	1.2
Season biomass of fungivores on Site I	450	270	350	250

[1]Density in numbers per square meter and biomass in milligrams per square meter.

Site I showed a significant ($P < 0.05$) interaction between treatment and information content measured by number of individuals, as well as between treatment and richness of aboveground arthropod species (Figure 6.12). However, number-based equitability did not exhibit a significant response to treatment. The trend for all three measures of diversity was similar, with the control plot having a relatively low value, the low-treatment plot the highest mean, and the moderate- and high-SO_2 exposures having successively lower means (Figure 6.12). Using biomass rather than individual numbers as a basis resulted in no significant interaction for treatment and either information content or equitability for these aboveground arthropods. The trend for the biomass measures was lowest in the control plot followed by few differences among the three exposure treatments.

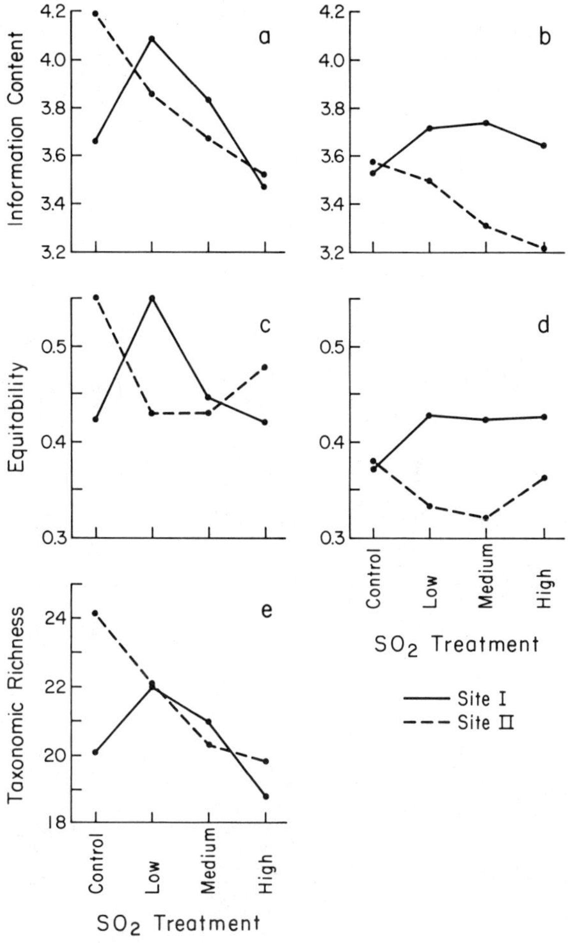

Figure 6.12. Information content for aboveground arthropods on the Site I and II SO_2 treatments measured using numbers (a) and biomass (b), equitability measured using numbers (c) and biomass (d), and taxonomic richness (e).

On Site II, the aboveground arthropod data showed no significant effects due to treatment in any of the calculated measures for diversity. Unlike Site I, diversity of the control plot was higher than for any of the treatments (Figure 6.12). The subsequent general trend was for a decline across treatments for both biomass and individual number-based information content and richness. However, equitability levels showed an increase on the highest SO_2 treatment.

Diversity showed no significant interaction with treatment for the soil microarthropod families on either Site I or II (Figure 6.13). Information content was a minimum on the low-treatment plots for all except the biomass-based estimate on Site II, where the high treatment had the minimum mean. Equitability showed a decline across treatments with a biomass base on Site I and a possible overall

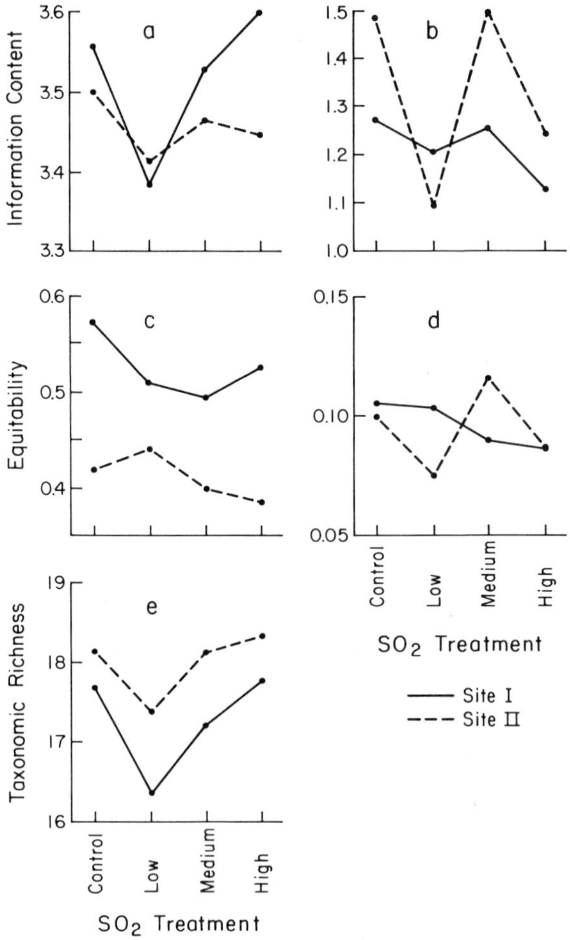

Figure 6.13. Information content for soil microarthropod families on the Site I and II SO_2 treatments measured using numbers (a) and biomass (b), equitability measured using numbers (c) and biomass (d), and taxonomic richness (e).

decline with a numbers base. The numbers-based decline was repeated for Site II. Richness appeared to show an increase across treatments on both sites after an initial decline from the control for the low treatment.

It is apparent that factors other than SO_2 exposure may be the critical influences on arthropod diversity across the experimental plots. The soil microarthropods on Site I showed responses among the diversity measurements, which appear to be complementary to those for the aboveground arthropods. Where one measure declined, the other showed trends for increases in measured diversity, with the changes having the same relative magnitude (Figures 6.12 and 6.13).

6.5 Vertebrate Consumers

Domestic and wild ruminants are ecologically and economically important in Great Plains grasslands. Although the 0.5-ha SO_2 treatment plots could not accommodate direct experimentation with large consumers, potential indirect effects can be considered by examining plant responses in conjunction with the grazing and nutritional characteristics of ruminants.

6.5.1 Ruminants

The energy and nitrogen balance of ruminants is a function of both voluntary intake and forage quality. Voluntary intake of forage is complexly related to interactions of availability, palatability, digestibility, and rate of passage of material through the digestive tract (Milchunas et al., 1978). These latter three components of voluntary intake are in turn correlated with forage quality characteristics, such as the proportions of structural to labile material, nutrient element content, the interactions of these nutrients in metabolic processes, and allelopathic or toxic components.

The high-sulfur concentrations observed in plants exposed to SO_2 (see Section 4.3.1) can potentially alter forage digestibility. Rumen microorganisms are capable of utilizing nitrogen and sulfur in the synthesis of amino acids (Abdo et al., 1964), and the digestion of cellulose by rumen microbes can be stimulated by either sulfur (Barton et al., 1971) or nitrogen (Milchunas et al., 1978) supplementation. On the other hand, sulfur has been reported to be toxic to rumen microorganisms at a concentration of 100 ppm from sodium sulfate (Hubbert et al., 1958) and at 30 ppm from sodium sulfite (Trenkle et al., 1958). In Chapter 4 we reported that SO_2 exposure had no effect on the nitrogen content of *A. smithii* (see Table 4.5, Section 4.3.1), but had a large influence on N:S ratios (see Table 4.6, Section 4.3.1). The elevated sulfur content and altered N:S ratios did not, however, influence the in vitro digestibility of the forages for ruminants at any time during the growing season (Table 6.5). The expected seasonal decline in digestibility was observed.

The proportions of structural to labile material in forage have important implications with respect to rate of digestion and ingesta passage (Van Soest,

Table 6.5. *In Vitro* digestible dry matter (%) of *Agropyron smithii* for Seven Dates, Four Treatments, Two sites, and 3 years of SO_2 Exposure[1]

Site and Treatment	Year	Date April	May	June	July	Aug. 6	Aug. 30	Sept.	Treatment mean[2]
Site I	1975								
Control		—	71.8	54.2	50.6	45.7	—	40.7	50.5[a]
Low		—	64.4	53.9	47.1	44.4	—	34.1	49.5[a]
Medium		—	69.2	55.5	50.7	44.2	—	39.4	49.9[a]
High		—	69.0	60.0	51.9	45.9	—	37.3	51.9[a]
Date mean		—	67.9[a]	55.9[b]	50.1[c]	45.1[d]	—	38.1[e]	50.4
Site I	1976								
Control		—	62.6	49.4	47.8	41.8	—	37.0	47.7[a]
Low		—	62.1	48.5	44.1	41.8	—	36.2	46.5[a]
Medium		—	60.0	50.5	44.4	40.8	—	37.3	46.6[a]
High		—	58.2	50.6	45.0	38.7	—	35.5	45.6[a]
Date mean		—	60.7[a]	49.8[b]	45.3[c]	40.8[d]	—	36.5[e]	46.6
Site I	1977								
Control		83.2	72.4	71.2	61.0	55.7	51.4	54.1	64.7[a]
Low		83.7	70.9	—	58.3	55.5	52.0	47.8	59.0[a]
Medium		83.6	69.0	—	58.0	53.1	53.7	53.1	61.1[a]
High		83.1	74.1	67.1	58.1	56.9	51.0	52.8	61.6[a]
Date mean		83.4[a]	71.5[b]	69.2[b]	58.9[c]	55.2[d]	52.1[d]	51.7[d]	61.8
Site II	1976								
Control		—	60.5	56.1	48.5	42.8	—	39.0	49.4[a]
Low		—	66.9	54.3	50.7	45.0	—	40.0	51.4[a]
Medium		—	66.7	52.7	50.1	43.3	—	35.0	49.6[a]
High		—	69.2	54.6	47.9	46.5	—	41.1	52.1[a]
Date mean		—	65.8[a]	54.4[b]	49.3[c]	44.4[d]	—	39.0[e]	50.6
Site II	1977								
Control		84.8	73.5	—	65.9	64.4	—	—	70.3[a]
Low		84.2	74.9	—	63.4	61.2	—	—	70.9[a]
Medium		82.3	71.4	—	64.5	62.0	—	—	69.2[a]
High		82.1	72.6	—	63.8	61.6	—	—	70.1[a]
Date mean		83.3[a]	73.1[b]	—	64.4[c]	62.3[c]	—	—	70.2

[1]Any two means not sharing a common superscript within a row, or column for a given site and year, are significantly different ($P < 0.05$). Adapted from Milchunas et al. (1981).
[2]For May, June, July, and August data.

1975). Cell wall content (primarily cellulose, hemicellulose, and lignin) of *A. smithii* was not altered by SO_2 treatment (Table 6.6).

SO_2–sulfur is converted to sufate salts in plant tissue (Thomas et al., 1951), and increases in calcium, magnesium, and potassium as a result of exposure to SO_2 have been reported (Ziegler, 1975). Mineral ash content of *A. smithii* increased with increasing SO_2 concentration (Table 6.6). The increased $SO_4^{=}$ content of plants exposed to SO_2 may increase their requirement for cations, but may also increase the elemental quality of forage to herbivores.

Table 6.6. Ash and Cell Wall Contents (%) in *Agropyron smithii* for Control, Low-, Medium-, and High-SO_2 Treatments for Site I—1975, Site I—1976, and Site II—1976, All Months Combined[1]

	Site I—1975	Site I—1976	Site II—1976
		Ash	
Control	7.4[a]	6.7[a]	8.4[b]
Low	7.5[a]	7.3[b]	7.6[a]
Medium	8.0[b]	7.4[b]	8.9[c]
High	8.3[b]	8.5[c]	9.0[c]
		Cell Wall Material[2]	
Control	68.1[a]	68.4[a]	63.0[a]
Low	68.9[a]	68.4[a]	63.9[a]
Medium	68.7[a]	68.2[a]	64.4[a]
High	67.5[a]	65.7[a]	60.6[a]

[1]Any two means within a constituent not sharing a common superscript within a column are significantly different ($P < 0.05$).
[2]Neutral detergent fiber (NDF).

The high-sulfur content in forage exposed to SO_2 can alter the nutritional or toxic potential of other elements because of metabolic interactions that occur in both the plant and the animal. Increased sulfur content of forages in spring can improve N:S ratios (see Table 4.6, Section 4.3.1) and thereby increase ruminant nitrogen retention (see Figure 7.12, Section 7.6.2). In other cases, negative or positive responses may result. For example, both requirement and toxic levels of selenium for ruminants and monogastrics are influenced by dietary sulfur levels (Whanger, 1970). Further, the relative proportions of sulfur to selenium in the soil can influence the uptake of these elements by roots and thereby alter forage quality. Sulfur–selenium interactions can be attributed to competitive antagonism between structural analogs (Harborne, 1977).

To assess possible effects of atmospheric and soil sulfur on forage selenium content, subplots on Site I were fertilized with magnesium sulfate and sodium selenate (Milchunas et al., 1982). Interactions of SO_2 with sulfur or selenium fertilizer were found mainly on the high-SO_2 treatment. A trend of decreased forage selenium content with SO_2 exposure was observed. Forage selenium concentrations averaged 2.1 $\mu g \cdot g^{-1}$ on the control and 1.8 $\mu g \cdot g^{-1}$ on the SO_2 treatments. High-soil selenium reduced the sulfur content of forage but high levels of sulfur in the soil did not decrease forage selenium content except when selenium-fertilized plants were compared to sulfur plus selenium-fertilized plants. Although competitive inhibition of root uptake was observed between the two elements when their concentrations were elevated by fertilization, naturally occurring selenium levels on the plots were perhaps too low to detect changes. With no fertilization, the mean S:Se ratio of forage was 2530 on the control and 15,370 on the high-SO_2 treatment. The sixfold dilution of selenium in forage could make it necessary to increase selenium supplementation in order to meet animal requirements. Although beyond the scope of this discussion, further assessments of the impact of SO_2 on ruminants should also include a consideration of the effect of dietary sulfur level on ruminant molybdenum metabolism (Grace and Suttle, 1979) and Cu/Mo/$SO_4^=$ interactions (Matrone, 1970).

The sulfur concentration of forages can positively and negatively affect voluntary intake. Sulfur fertilization that increased pangola grass sulfur concentration from 0.09 to 0.15% increased voluntary intake by sheep from 44.4 to 64.1 $g \cdot kg\ W^{0.75} \cdot day^{-1}$ (Rees et al., 1974), and this was probably due to the concurrent 5% increase in digestibility. On the other hand, high dietary sulfur concentrations can reduce voluntary intake (Rumsey, 1978; Bouchard and Conrad, 1974), and this can be attributed to reduced palatability because the digestibility of nitrogen, sulfur, and dry matter remained unchanged. Although we do not know whether sulfur concentrations of forage on our SO_2 treatments were high enough to cause reductions in voluntary intake by ruminants, we did observe a decline in palatability of *A. smithii* to grasshoppers (Section 6.3.3).

The structure of the plant canopy can influence the availability of forage to consumers and the quality of forage that is consumed. The structural aspects of concern in this case include the distribution in time and space of live versus dead material and fibrous versus soluble plant components. Live plant tissue contains more protein, soluble carbohydrates, and other nutrients than dead plant tissue. Grazing animals show a high preference for green material (Arnold, 1962; Reppert, 1960). However, when green leaves are mixed with dead leaves, intake by consumers may decrease dramatically (Arnold, 1962). Further, total dry matter intake decreases as the proportion of dead material in the diet increases because of increased rumen ballast and decreased rates of digestion and passage (Demarquilly et al., 1965). The vertical distribution of the live and dead plant components is important because, although grazing animals move in a horizontal plane, they select in a vertical plane (Arnold, 1962).

Canopy leaf area profiles (see Figures 5.1 and 5.2, Section 5.2.1) and live-to-dead ratios indicated that the SO_2 treatments may have an effect on the availability of forage to consumers. In June, total live leaf area was greater on the control and the live-to-dead ratio was greatest on the low-SO_2 treatment. Through the rest of the growing season, the low-SO_2 treatment was structurally optimum from a grazing standpoint in leaf area production, live-to-dead ratios, and the distribution of the live compared with the dead material. This optimum declined progressively from the low, medium-, high-, to the control SO_2 treatments.

Nitrogen concentrations, *in vitro* digestibilities and the proportion of structural components to cell-soluble components were not affected by the SO_2 treatments. Alterations in the availability of forage to ruminants and the quality of the forage appear to be related to the influence of SO_2 on structural aspects of the plant community rather than a direct effect on forage quality.

6.6 Summary

The effects of the SO_2 treatments on soil heterotrophs were primarily a function of the depth in the soil the organisms occupied and their seasonal activity pattern.

Surface-dwelling tardigrades were the most noticeably impacted group of invertebrates. Non-stylet-bearing nematode densities decreased in the top 10 cm of soil, but were not affected in the 10- to 20-cm depth.

The effects of SO_2 on soil microarthropod populations appeared to be related to activity. Populations were reduced with exposure to SO_2 during the early part of the growing season when high-soil water supported large, active populations. The effects of SO_2 diminished as the soil dried and activity declined through the growing season. Some populations did, however, increase with SO_2 exposure. Overall net changes in population biomass or density were not observed.

With respect to aboveground macroarthropods, major families, and therefore total biomass and density, were not affected by the SO_2 treatments although numerous, less abundant taxa were affected. When categorized by trophic groups, parasite–parasitoid and plant sucking aboveground arthropods decreased with SO_2 exposure and pollen–nectar feeders displayed a subsidy–stress response gradient, i.e., increased on the low treatment and then declined below control levels with additional SO_2 exposure. Grasshopper densities declined with increasing SO_2 exposure concentration, but no effect on species composition could be detected. The decline in grasshopper density with SO_2 treatment could be attriubted to both toxicity and avoidance mechanisms because nymph mortality, egg densities, and forage palatability were reduced with SO_2 exposure.

The effects of the SO_2 on ruminant consumers based on indices of forage quality, i.e., *in vitro* digestibility, nitrogen content, forage fiber to soluble components, did not indicate any treatment effects. Increased concentrations of sulfur in plants exposed to SO_2 can affect consumers by influencing the availability, requirements, and metabolism of other minerals. Alterations in the availability of forage to grazing animals and the quality of the forage appeared to be primarily related to the influence of SO_2 on structural aspects of the plant community rather than a direct effect on forage quality. Based on plant community structure data presented in Chapter 5, a subsidy–stress gradient across low-, medium-, and high-SO_2 treatments may best describe the overall effect on ruminants.

References

Abdo, K. M., K. W. King, and R. W. Engel. 1964. Protein quality of rumen microorganisms. *J. Anim. Sci.* 23:734–739.

Arnold, G. W. 1962. Factors within plant associations affecting the behavior and performance of grazing animals. In *Grazing in Terrestrial and Marine Environments*, D. J. Crisp, ed. *Br. Ecol. Soc. Symp.* 4:133–154.

Barton, J. S., L. S. Bull, and R. W. Henten. 1971. Effects of various levels of sulfur upon cellulose digestion in purified diets and lignocellulose digestion in corn fodder pellets *in vitro. J. Anim. Sci.* 33:682–685.

Bouchard, R., and H. R. Conrad. 1974. Sulfur metabolism and nutritional changes in lactating cows associated with supplemental sulfate and methionine hydroxy analog. *Can. J. Anim. Sci.* 54:587–593.

Bromenshenk, J. J., and C. C. Gordon. 1978. Terrestrial insects sense air pollutants. In *Conference Proceedings, First Joint Conference on the Sensing of Environmental Pollutants, 1977.* pp. 66–70. Washington, D.C.: American Chemical Society.

Cates, R. G., and G. H. Orians. 1975. Successional status and the palatability of plants to generalized herbivores. *Ecology* 56:410–418.

Demarquilly, C., J. M. Borssau, and G. Cuylle. 1965. Factors affecting the voluntary intake of green forage by sheep. São Paulo, Brazil: *Proceedings of the IX International Grassland Congress*. pp. 877–885.

Dodd, J. L., and W. K. Lauenroth. 1981. Effects of low-level SO_2 fumigation on decomposition of western wheatgrass litter in a mixed grass prairie. *Water Air Soil Pollut.* 15:257–261.

Ginevan, M. E., and D. D. Lane. 1978. Effects of sulfur dioxide in air on the fruit fly, *Dryosphila melanogaster. Environ. Sci. Technol.* 12:828–831.

Grace, N. D., and N. F. Suttle. 1979. Some effects of sulfur intake on molybdenum metaoblism in sheep. *Br. J. Nutr.* 41:125–136.

Harborne, J. B. 1977. *Introduction to Ecological Biochemistry*. London: Academic Press.

Hillman, R. C., and A. W. Benton. 1972. Biological effects of air pollution on insects, emphasizing the reactions of the honey bee (*Apis mellifera* L.) to Sulfur Dioxide. *Elisa Mitchell Sci. Soc.* 88:195.

Hubbert, F., Jr., E. Cheng, and W. Burroughs. 1958. Mineral requirements of rumen microorganisms for cellulose digestion *in vitro. J. Anim. Sci.* 17:559–568.

Hurlbert, S. H. 1971. The nonconcept of species diversity: A critique and alternative parameters. *Ecology* 52:577–586.

Kitchell, J. F., R. V. O'Neill, D. Webb, G. W. Gallepp, S. M. Bartell, J. F. Koonce, and B. S. Ausmus. 1979. Consumer regulation of nutrient cycling. *BioScience* 29:28–34.

Lebrun, Ph. J. M. Jacques, M. Goossens, and G. Wauthy. 1978. The effect of interaction between the concentration of SO_2 and the relative humidity of air on the survival of the bark-living bioindicator mite *Humerobates rostrolamellatus. Water Air Soil Pollut.* 10:269–275.

Lee, J. J., and D. L. Inman. 1975. The ecological role of consumers—An aggregated systems view. *Ecology* 56:1455–1458.

Leetham, J. W., J. L. Dodd, J. S. Logan, and W. K. Lauenroth. 1980. Response of *Melanoplus sanguinipes* to low-level sulfur dioxide exposure from egg hatch to adult (Orthoptera:Acrididae). In *The Bioenvironmental Impact of a Coal-Fired Power Plant*, E. M. Preston, D. W. O'Guinn, and R. A. Wilson, eds. pp. 176–184. Sixth Interim Report, Colstrip, Montana. Corvallis, Oregon: EPA-600/3-81-007.

Leetham, J. W., T. J. McNary, J. L. Dodd, and W. K. Lauenroth. 1981a. Response of soil nematodes, rotifers and tardigrades to three levels of season-long sulfur dioxide exposures. *Water Air Soil Pollut.* 17:343–356.

Leetham, J. W., J. L. Dodd, R. D. Deblinger, and W. K. Lauenroth. 1981b. Arthropod population responses to three levels of chronic sulfur dioxide exposure in a northern mixed-grass ecosystem. I. Soil microarthropods. In *The Bioenvironmental Impact of a Coal-Fired Power Plant*, E. M. Preston, D. W. O'Guinn, and R. A. Wilson, eds. pp. 139–157. Sixth Interim Report, Colstrip, Montana. Corvallis, Oregon: EPA-600/3-81-007.

Leetham, J. W., J. L. Dodd, R. D. Deblinger, and W. K. Lauenroth. 1981c. Anthropod population responses to three levels of chronic sulfur dioxide exposure in a northern mixed-grass ecosystem. II. Aboveground arthropods. In *The Bioenvironmental Impact of a Coal-Fired Power Plant*, E. M. Preston, D. W. O'Guinn, and R. A. Wilson, eds. pp. 158–175. Sixth Interim Report, Colstrip, Montana. Corvallis, Oregon: EPA-600/3-81-007.

Leetham, J. W., J. L. Dodd, and W. K. Lauenroth. 1983. Effects of low-level sulfur dioxide exposure on decomposition of *Agropyron smithii* litter under laboratory conditions. *Water, Air, Soil Pollut.* 19:247–250.

Lloyd, M., and R. J. Ghelardi. 1964. A table for calculating the equitability component of species diversity. *J. Anim. Ecol.* 33:217–225.

Matrone, G. 1970. Studies on copper-molybdenum-sulphate interrelationship. In *Trace Element Metabolism in Animals*, C. F. Mills, ed. pp. 354–361. London: E. & S. Livingstone.

Mattson, W. J., and N. D. Addy. 1975. Phytophagous insects as regulators of forest primary production. *Science* 190:515–522.

McNary, T. J., D. G. Milchunas, J. W. Leetham, W. K. Lauenroth, and J. L. Dodd. 1981. Effects of controlled low levels of SO_2 on grasshopper densities on a northern mixed-grass prairie. *J. Econ. Entomol.* 74:91–93.

Milchunas, D. G., M. I. Dyer, O. C. Wallmo, and D. E. Johnson. 1978. *In-vivo/in-vitro* relationships of Colorado mule deer forages. *Colo. Div. Wildl.* Special Rep. No. 43.

Milchunas, D. G., W. K. Lauenroth, and J. L. Dodd. 1981. Forage quality of western wheatgrass exposed to sulfur dioxide. *J. Range Manage.* 34:282–285.

Milchunas, D. G., W. K. Lauenroth, and J. L. Dodd. 1983. The interaction of atmospheric and soil sulfur on the sulfur and selenium concentration of range plants. *Plant Soil* 72:117–125.

Mitchell, J. E., and R. E. Pfadt. 1974. A role of grasshoppers in a short-grass prairie ecosystem. *Environ. Entomol.* 3:358–360.

O'Neill, R. V. 1976. Ecosystem persistence and heterotrophic regulation. *Ecology* 57:1244–1253.

Peet, P. K. 1974. The measurement of species diversity. *Annu. Rev. Ecol. Syst.* 5:285–306.

Pielou, E. C. 1966. The measurement of diversity in different types of biological collections. *J. Theor. Biol.* 13:131–144.

Rees, M. C., D. J. Minson, and F. W. Smith. 1974. The effect of supplementary and fertilizer sulphur on voluntary intake, digestibility, retention time in the rumen, and site of digestion of pangola grass in sheep. *J. Agric. Sci. (Camb.)* 82:419–422.

Reppert, J. N. 1960. Forage preference and grazing habits of cattle at the Eastern Colorado Range Station. *J. Range Manage.* 13:58–65.

Rumsey, T. S. 1978. Effects of dietary sulfur addition and synovex-S ear implants on feedlot steers fed an all-concentrate finishing diet. *J. Anim. Sci.* 46:463–477.

Thomas, W. E., J. K. Loosli, H. H. Williams, and L. A. Maynard. 1951. The utilization of inorganic sulfates and urea nitrogen by lambs. *J. Nutr.* 43:515–523.

Trenkle, A., E. Cheng, and W. Burroughs. 1958. Availability of different sulfur sources for rumen micro-organisms in *in vitro* cellulose digestion. *J. Anim. Sci.* 17:1191.

Van Soest, P. J. 1975. Physico-chemical aspects of fibre digestion. In *Digestion and Metabolism in the Ruminant*, I. W. McDonald and A. C. I. Warner, eds. pp. 352–365. Proc. IV International Symp. Ruminant Physiol. Sydney, Australia.

Wallwork, J. A. 1970. *Ecology of Soil Animals*. Maidenhead, Berkshire, England: McGraw–Hill.

Whanger, P. D. 1970. Sulphur-selenium relationships in animal nutrition. *Sulphur Inst. J.* 6:6–10.

Ziegler, I. 1975. The effects of SO_2 pollution on plant metabolism. *Residue Rev.* 56:79–105.

7. Simulation of SO_2 Impacts

J. E. HEASLEY, W. K. LAUENROTH, AND T. P. YORKS

7.1 Introduction

Mathematical modeling is a long-standing tool for the analysis of ecological systems (Lotka, 1925). Before the elaboration of systems engineering, beginning during World War II, applications of mathematical models in ecology were limited to small systems (few equations). The widespread availability of computers in the late 1950s and the 1960s made it possible for ecologists to explore the potential of systems analysis to contribute to the study of ecological systems. This period of experimenting with systems analysis techniques coincided with a period during which many ecologists were switching their interests from the study of populations and communities to ecosystems (Odum, 1957; Margalef, 1963). Systems analysis, computers, and ecological systems were merged into systems ecology in the mid-1960s (Odum, 1960; Olson, 1963; Patten and Witkamp, 1967; Van Dyne, 1969; Watt, 1966).

The major difference between the previous applications of mathematical models in ecology and the approach of systems ecology centered on the complexity of the problems addressed. Systems analysis techniques and computers enabled ecologists to develop a total system approach to ecological problems (Van Dyne, 1969). Large systems of equations could be solved numerically. This provided freedom from the major limitations of early approaches; a small number of equations and linearity.

The impact of the introduction of systems analysis into ecology in the 1970s is best represented by the central role of modeling in the International Biological Program (IBP) projects in the United States (Patten, 1975). The Grassland Biome project at Colorado State University substantially influenced grassland and rangeland ecology in the United States (Van Dyne, 1972). The systems analysis approach developed during the Grassland Biome project was directed toward modeling total ecosystems at a relatvely high level of detail (Innis, 1975).

A major product of the Grassland Biome project modeling activities was a total-system model called ELM (Innis, 1978). This model was constructed by a team of scientists, contained approximately 120 state variables, and represented 20 to 30 person-years of effort (Innis, 1975; Innis et al., 1980; Van Dyne, 1978). This model was constructed with a mechanistic philosophy which leads to reduction of processes to basic physicochemical laws or, failing this, attempts to isolate causal pathways (Innis et al., 1980). Simulation modeling activities reported here were conducted utilizing the mechanistic philosophy and took the ELM model as the starting point.

This chapter begins with a brief description of the submodels that comprise the simulation model, including a discussion of the implementation of the effects of SO_2. Model sensitivity and results of validation tests are discussed next. Experiments conducted with the model are described, and the results are discussed and compared with the findings from the field experiments.

7.2 Model Description

Our objective for constructing a simulation model was to provide a means of evaluating system-level impacts as a result of exposing a northern mixed prairie to SO_2. The grassland model (SAGE—Systems Analysis of Grassland Systems) is a difference-equation, flow-oriented, simulation model (Heasley et al., 1981). The ELM grassland simulation model (Innis, 1978) was used as a starting point. Substantial restructuring and rewriting resulted in SAGE being a different model rather than simply a refinement of ELM. This model simulates the flow of carbon, nitrogen, and sulfur through the major components of a northern mixed-prairie system (Figure 7.1). These components are

1. Abiotic subsystem
2. Primary producer subsystem
3. Soil subsystem
4. Ruminant consumer subsystem

The primary driving variables are solar radiation, air temperature, precipitation, relative humidity, and wind speed. The model simulates the response of a representative square meter of grassland. The difference equation representations of processes are implemented in FORTRAN utilizing MODAID, a collection of simulation programs (Kirchner and Vevea, 1983). The general time resolution for the model is 1 day with a few processes being simulated within the day.

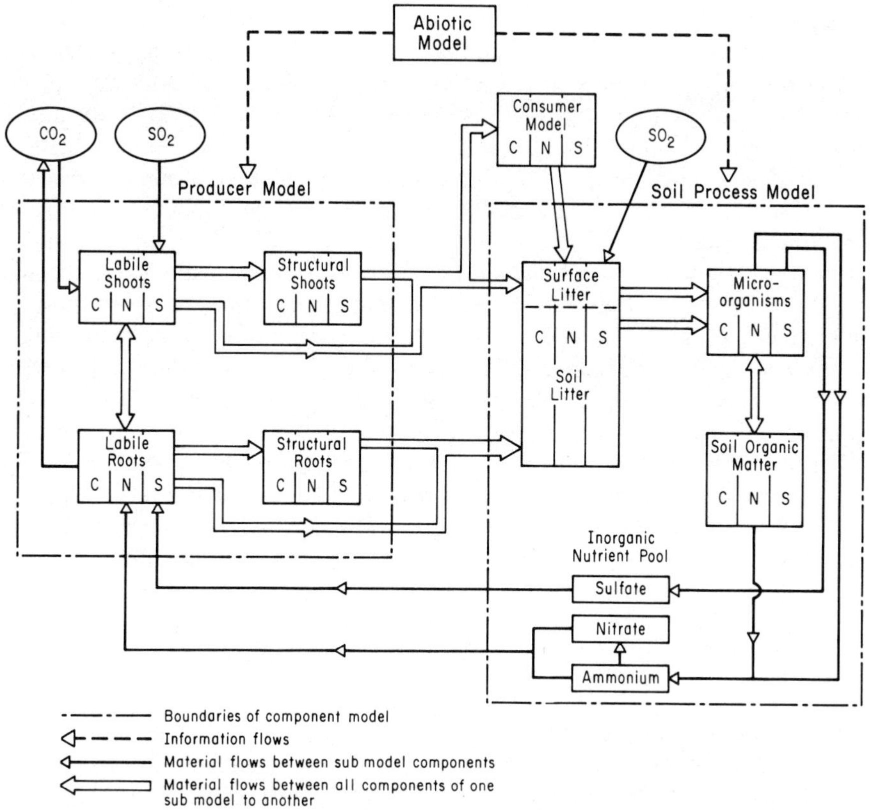

Figure 7.1. System diagram of SAGE a grassland model.

Atmospheric SO_2 enters the system by means of an SO_2 deposition submodel. The effects of the SO_2 are simulated at the process level.

7.2.1 Abiotic Submodel

The abiotic model was designed to simulate the soil and canopy abiotic variables which influence a grassland, and consists of a water flow submodel and a temperature profile submodel (Parton, 1978). The water flow submodel simulates the flow of water through the plant canopy and several soil layers. The partitioning of precipitation into evaporation and transpiration is the most important of these processes. The model is generalized to handle a number of layers of soil water of specified depth and soil type and is structured to include the important feedback mechanisms between the biotic and abiotic state variables.

The temperature profile submodel simulates daily solar radiation, maximum canopy air temperature, and soil temperature at 13 points in the soil profile. Soil temperature is calculated with a modified finite difference solution of the one-dimensional Fourier heat conduction equation when temperature at the upper (0–cm) and lower (180–cm) interfaces are specified. Temperature at the upper interface is calculated as a function of the air temperature at 2 m, the potential evapotranspiration rate, and the standing crop biomass. The monthly average temperature at 180 cm is used in calculating the lower interface. Solar radiation is simulated as a function of cloud cover and the time of year, while the maximum canopy air temperature is a function of solar radiation and the maximum air temperature at 2 m.

The water flow submodel calculates the flow of water through the plant canopy into the soil water layers. Water loss by standing crop and litter interception are subtracted from daily rainfall, with the remaining rainfall infiltrated into the soil. Water loss via the interception of rainfall by litter and the standing crop is due to evaporation, while water loss from the soil is a function of bare soil evaporation and transpiration.

The temperature profile and water flow submodels use daily precipitation, cloud cover, wind speed (2 m), maximum and minimum air temperatures (2 m), and relative humidity (2 m) as driving variables. These variables can be determined either by using an observed time series of daily weather observations or by a stochastic weather simulator.

7.2.2 Primary Producer Submodel

The primary producer submodel simulates the carbon (C), nitrogen (N), and sulfur (S) dynamics of three groups of grassland plants, cool-season grasses, warm-season grasses, and cool-season forbs (Table 2.5). Both structural and labile forms of C, N, and S are represented. Aboveground plant parts are divided into young tissue (actively growing) and mature tissue (nongrowing but photosynthetically active). Belowground plant parts are separated into crowns, rhizomes, and roots. Roots are further distinguished according to four soil layers: 0–1, 1–5, 5–20, and 20–60 cm.

Carbon enters the primary producer subsystem via photosynthesis. This process is represented by a CO_2 diffusion model, which combines a leaf diffusion resistance network with a double Michaelis–Menton representation of carboxylation. Carbohydrate production responds to leaf temperature, atmospheric CO_2 concentration, leaf water potential, leaf age, leaf temperature acclimation, and light intensity. The distinction between the C_3 and the C_4 photosynthetic pathways of the three plant categories is accomplished by parameter values.

In the absence of air pollutants, nitrogen and sulfur enter the primary producer subsystem via root uptake from soil nutrient pools. Sulfur is taken up as sulfate and nitrogen is taken up as ammonium and nitrate. The rate of root uptake is determined by the soil solution concentrations of sulfate, ammonium, or nitrate (Michaelis–Menton). This basic form is modified by the effects of soil temperature, soil moisture, and the relative levels of labile sulfur or nitrogen in the roots.

Growth of roots and shoots is represented as an exponential function of young tissue carbon. This function is in turn modified by temperature and moisture.

Structural tissue is assumed to be synthesized with fixed C:N and C:S ratios. The labile elements associated with structural material are also assumed to occur with fixed C:N and C:S ratios. Therefore, carbon growth requirements dictate nitrogen and sulfur growth requirements. If either nitrogen or sulfur levels in the plant are less than required, carbon growth is reduced accordingly.

Nutrient allocation is a function of the demands for carbon, nitrogen, and sulfur, the supply of carbon, nitrogen, and sulfur, and the age structure of the plant tissue. Young tissue is assumed to have labile material associated with it to support growth. As young tissue approaches maturity, the amount of labile support material approaches zero. Therefore, the sulfur and nitrogen concentration of the plant reflects the age structure of the tissue in that plant. Nutrients are allocated via a priority scheme with maintenance functions (respiration) being met first, followed by growth and reproduction. Shoot requirements that cannot be satisfied by nutrient levels contained in the shoots may be met through translocation from belowground organs. The amount that can be translocated depends upon the relative amounts of labile nutrients contained in the crowns and roots and the maximum translocation rates from roots to shoots. Excess shoot nutrients may be translocated to belowground plant parts subject to the constraints placed upon the translocation process (relative labile levels and translocation rates).

Carbon, sulfur, and nitrogen leave the primary producer subsystem via respiration, root exudation, senescence, litter fall, and consumption by ruminants. Respiration of roots and shoots is simulated as an exponential function of carbon levels, temperature, and moisture. Root exudation represents the leakage of nutrients across the membranes into the soil pools. This is primarily a function of the concentration of labile carbon, nitrogen, and sulfur contained in the roots. Senescence is simulated as a function of tissue age, temperature, and soil moisture levels. Standing dead material enters the soil process subsystem via litter fall. Litter fall is estimated as a function of precipitation and standing dead carbon.

7.2.3 Soil Process Submodel

The soil submodel simulates processes restricted to the soil subsystem (McGill et al., 1981). These include inorganic nutrient transformations, litter decomposition (above- and belowground), microbial processes (uptake, growth, death), fractionation of soil organic matter into humic (humads) and recalcitrant forms, and the movement of nutrients between soil layers (leaching). Two forms of inorganic nitrogen (ammonium and nitrate) and one form of inorganic sulfur (sulfate) are simulated. In the majority of processes, carbon, nitrogen, and sulfur move through the soil subsystem in a parallel fashion (Figure 7.1). When either plants or microbes die, they are partitioned into slowly decomposing low-N-content substrate (mainly cell walls) and rapidly decomposing, high-N-content substrate (mainly cytoplasm and organelles). The rate of use of these substrates by bacteria and fungi depends on the physiological state of the microbes, as reflected by their C:N, C:S, N:S ratios, and on soil temperature and moisture. Humads are fairly resistant materials formed directly from high-N substrate by adsorption onto humic materials and clay minerals, or left as a residue of resistant material, such as lignins, after the decomposition of low-N substrate. Resistant soil organic matter is the material left after the decomposition of humads. It is important to recognize these classes of substrate because their relative abundance determines the nutritive status of microbes, and whether microbes mineralize or immobilize nitrogen (McGill et al., 1981). Bacteria and fungi are distinguished in the model because they differ in their responses to abiotic factors, and in their abilities to break down the classes of substrate. Usually microbes are net mineralizers of nitrogen, but they may compete with plants for inorganic nitrogen if low-N substrate is abundant.

7.2.4 Ruminant Consumer Submodel

The ruminant consumer submodel simulates the energy, nitrogen, and sulfur requirements of ruminant grazers; their consumption and metabolism of energy, nitrogen, and sulfur; consequent periods of positive and negative energy and nitrogen balance; and their resultant weight gains or losses (Swift, 1983). The ruminant submodel consists of three parts: nitrogen and sulfur ingestion, metabolism, and loss; energy ingestion, metabolism, and loss; and a representation of ruminal microbial protein. The fate of sulfur in the ruminant's body is tied to the fate of nitrogen because both are involved primarily in protein metabolism.

This submodel is a straightforward representation of ruminant energy and nitrogen balance, but a few points are worthy of separate mention. Ingestion of forage by the animal is regulated by physical capacity of the rumen and the rate of passage of ingesta through the digestive tract. Passage rate is controlled by forage digestibility and rumen microbial biomass. The rate of microbial protein synthesis in the rumen is controlled in turn by the amount of energy released by digestion during the previous time step and by the amount of nitrogen in the rumen. A capability for recycling nonprotein nitrogen to the rumen from the blood exists and is triggered by low dietary nitrogen concentration.

Energy and nitrogen are metabolized jointly in the model. This permits the nitrogen status of the animal to influence the manner in which energy is partitioned and vice versa. The pool of labile energy in the body consists of amino acids as well as nonnitrogenous energy-bearing compounds contributed by rumen fermentation and by a normal mobilization of a portion of the fat stores (Koong and Lucas, 1973). The amount of energy in this pool, the proportion of amino acids in it, and the energetic and growth demands placed on it control the partitioning of the energy pool and the fate of the nitrogenous compounds. Energy is allocated to reproduction, lean body growth, and fattening, in that order, provided the nitrogenous component of the energy pool is not limiting.

Although sulfur flows parallel those of nitrogen, there are three departures from this stoichiometric scheme. They are (1) recycling of nonprotein sulfur (NPS) to and from the rumen, which is modeled independently from NPN recycling; (2) synthesis of microbial protein in which both N and S concentrations control synthesis rates; and (3) adjustments to the amino acid pool to account for the fact that absorbed amino acids have a biological value of less than 1.

Recycling of nonprotein sulfur is predicted from dietary sulfur concentration. If no addition to the rumen is predicted, it is drawing from the animal's nonprotein sulfur (NPS) pool. If a net loss is indicated, a flow from ruminal sulfur to the NPS pool occurs. The nitrogen to sulfur ratio controls the efficiency of microbial protein production in terms of forage nitrogen per kilocalorie digested. Adjustments are made in the amino acid N pool and amino acid S pool via deamination and desulfation such that the ratio of N pool to S pool corresponds to the N:S ratio of lean body tissue.

7.2.5 SO_2 Deposition and Effects

The deposition of sulfur dioxide is simulated utilizing a diffusion resistance scheme. Wind speed is predicted within the canopy by utilizing the wind speed at reference height and the equations of Ripley and Redmann (1976). The canopy is divided into six layers in which the aerodynamic and boundary layer resistances are predicted for the midpoint of each layer. The resistance to diffusion of SO_2 is represented by a parallel resistance network from the reference height to the soil surface. Diffusion of SO_2 branches at the midpoint of each canopy layer. Resistance to diffusion of SO_2 from the midpoint of each layer to the leaf surface or the interior of the leaf consists of a parallel/series combination of stomatal resistance, leaf surface resistance, and boundary layer resistance. The total flux of SO_2 to the canopy is determined by dividing the SO_2 concentration at the reference height by the total diffusion resistance from the soil surface up to the reference height. This is calculated by the parallel combination of all of the resistances. Total flux is then split among the canopy layers utilizing electrical analog computation techiques. This method allows the prediction of SO_2 concentration at the midpoint of each of the canopy layers. The uptake of SO_2 in each of the layers is estimated as a function of the flux into that layer and the total leaf area contained in it. It is therefore necessary to provide estimates of canopy height and leaf area distribution.

Sulfur dioxide enters the leaf sulfur pool by diffusion through the stomates. In the leaf, SO_2 is converted to sulfite which in turn is converted to sulfate. It is the rate of conversion from SO_2 to sulfate that controls the entry rate of atmospheric sulfur into the leaf sulfur pool. Any sulfur that enters the leaf sulfur pool via the stomates and is converted to sulfate is immediately available for leaf growth. Excess sulfur in the leaf sulfur pool may be translocated to the roots depending upon the relative levels of the shoot and root sulfur pool. If SO_2 enters the leaf at a rate greater than the rate of conversion of sulfite to sulfate, sulfite will built up in the leaf. The destruction of leaf tissue is directly related to the level of sulfite in the leaf.

Sulfur interacts directly and indirectly with various physiological processes in the plant. These include stomatal behavior, enzyme production, phenology, carbohydrate storage, and respiration. Exposure to very low levels of SO_2 is assumed to stimulate stomatal opening while higher levels are assumed to stimulate stomatal closure (Majernik and Mansfield, 1971; Unsworth et al., 1972; Schramel, 1975; Black and Unsworth, 1980). The degree of stimulation or depression of stomatal opening is dependent upon the vapor pressure deficit.

Sulfur may influence photosynthesis through the stimulation of enzyme production or the inhibition of photosynthetic enzymes. Various researchers have shown stimulatory effects of SO_2 at low levels on photosynthesis beyond those effects attributed to stomatal opening (White et al., 1974; Bennett and Hill, 1973; Koziol and Jordan, 1978). This effect is implemented in the model through increases in the enzyme levels which drive the carboxylation reactions in the CO_2 diffusion model. At higher concentrations, SO_2 has been reported to inhibit RuDP carboxylase, a primary photosynthetic enzyme (Ziegler, 1975). Inhibition of photosynthesis is implemented through reduction in the relative enzyme level utilized in the CO_2 diffusion model. Leaf respiration is increased in direct proportion to the amount of SO_2 which enters the leaf (Guderian and Van Haut, 1970). The effects of sulfur on plant phenology and plant carbohydrate storage are implemented indirectly through its effect on carbohydrate production.

Decomposition rates are influenced by sulfur dioxide (Table 4.9) (Dodd and Lauenroth, 1981, Leetham et al., 1983). The uptake of carbon from surface litter by microbes is reduced as a function of SO_2 concentration at the soil surface. In addition to direct effects, the current model structure allows the investigation of the indirect effects of SO_2 on phenological patterns (e.g., senescence), overall decomposition rates, the cycling rates of sulfur through the ecosystem, changes in carbohydrate storage levels, forage quality, primary and secondary productivity, and the relative abundance of plant groups.

7.3 Model Sensitivity

The SAGE model is very large and complex. Hence, formal sensitivity analysis, which relates changes in parameters to resulting changes in output variables, would be enormously time-consuming and expensive. Nevertheless, the timing, slope, and magnitude of output dynamics may be significantly altered by errors in the conceptualization or estimation of parameters or relationships. As a compromise between the ideal and the practical, we attempted to identify the parameters that were most critical to the character of the various output variables, and to portray sensitivity in a qualitative sense.

The parameters or relationships that influence the driving variables (soil, water, soil and air temperature, etc.) have particular significance to model output. The most critical are the relationships that describe evaporation and transpiration. The rate of evapotranspiration directly influences the soil water potential which, in turn, controls many physiological functions in grasslands. Direct measurements of the parameters in this relationship were not available; consequently, the parameters were adjusted to biomass and precipitation data.

The producer submodel contains the largest number of parameters or relationships critical to primary output variables. One of the most sensitive is the effect of soil water potential on growth. A 10 to 20% change in the soil water potential at which shoot growth stops results in a shift in the timing of peak biomass by several days. The magnitude of peak biomass follows in the direction of that shift, increasing or decreasing accordingly. Field data were not available for the parameters in the relationship. In this case the parameters were adjusted so that the timing of peak biomass agreed with field data.

Shoot growth was initially assumed to be exponential, but field data showed that the growth rate decreased as the growing season progressed, and that the decline occurred when water was not limiting. We assumed that this decrease was caused by nutrient limitation, particularly of nitrogen, and therefore that the ability of plants to take up nitrogen played an important role in determining the shape of the growth side of the biomass curve. The maximum rate of nitrogen uptake by roots was modified to give a good fit of model output to biomass data during the most active portion of the growing season. This parameter also influences the magnitude of peak biomass. Soil moisture was critical to shoot senescence rate which, in turn, controls the shape of the live biomass curve during the senescence period of the growing seasion.

Root dynamics are very difficult to measure in the field. The model does not, therefore, contain any feedback controls such as root crowding or spatial variability of the soil to check the expansion of the root system. Field data did allow us to conclude, however, that live plus dead roots remain fairly constant in the normal situation. Accordingly, we assumed that the root system was in steady state, neither expanding nor contracting. We adjusted parameters that control root dynamics such as root respiration, growth, death, and exudation rates, so that total roots remained fairly constant. Errors in parameters or relationships could result in an unstable root system, leading to extinction of the vegetation or the intermediate step of having all the biomass tied up underground.

The fraction of carbon that is structural or labile in plants is important to the distribution of nutrients and carbohydrates in the producer subsystem through control of translocation and allocation of nutrients among plant parts. Similarly, parameters controlling carbohydrate storage and the use of those stores influences the survival of the plants. If insufficient carbohydrates are available for growth initiation early in the growing season, the growth cycle cannot begin. During short periods of severe stress, respiration rates must be curtailed to preclude exhaustion of the carbohydrate reserves. Changes in these parameters and relationships will directly change the survival capabilities of the plants. Values for these parameters were derived through model experimentation.

Microorganism dynamics strongly affect the dynamics of litter and soil nutrient pools. Parameters controlling microbial death, litter uptake, and mineralization are important to changes in soil litter and nutrient–pool levels. These parameters were set so that litter and nutrient-pool levels fluctuated only a small amount from 1 year to the next at the beginning of the year, but may vary considerably within any given year.

In the ruminant submodel, the most sensitive parameter is the maximum rumen capacity. The rumen capacity strongly affects whether an animal gains or loses weight, as well as the magnitude of that gain or loss. The value for this parameter was adjusted to give average growth patterns for a normal weather year.

For the SO_2 deposition submodel, wind speed and how it changes in the plant canopy are the most important relationships in the accumulation of sulfur in the system. Sulfur dioxide flux is directly dependent on wind speed, but stomatal resistance is important in controlling entry into the plants. The rate of conversion of sulfite to sulfate in the plants determines the amount of sulfur that is available for translocation to the roots, and thereby determines the relative shoot–sulfur concentration. The field data indicated that as sulfur accumulates in the leaves, the rate of accumulation declined. This feedback control is crucial in limiting the amount of sulfur accumulated over a growing season, and the parameters for this relationship were adjusted so that model sulfur concentrations agreed with field data for the several SO_2 exposure levels.

Each of the parameters and relationships described is sensitive by itself. The degree of sensitivity is, however, dependent on the state of the entire system and on the magnitude of the compensating feedbacks. As a result, the parameters are most critical during periods of limiting resources. Some of the parameters are interdependent, and must be adjusted in concert. Finally, as must be obvious in a model of this size, there may be many more sensitive parameters which would be revealed by more intensive testing.

7.4 Model Validation

Model validation has as its basic purpose the creation of confidence in users that the behavior described by the model is sufficiently realistic to satisfy their objectives in using it. In the construction of SAGE, subsystem-level processes were constructed in modules and tested against data when available. When data were not available, the modules were tested over a wide range of inputs to ensure that unrealistic responses would be detected and their causes corrected. Such tests were conducted for photosynthesis, nutrient uptake, nutrient allocation, growth, translocation, respiration, senescence, litter fall, surface litter dynamics, microbial dynamics, ruminant body weight, soil water content, and SO_2 deposition. After integrating the various process-level modules to form the system-level model, more tests were conducted using field data to check the overall model utility through such system-level outputs as biomass and nutrient concentrations. Parameter adjustments were made as deemed necessary. Validation concluded with comparison against field data without further parameter adjustment.

Three years of experimental data were utilized to make the adjustments in parameters and to compare model output with field data. Data collected during the 1975 growing season were used to make adjustments in parameters while 1976 and 1977 data were utilized for validation. The model was initialized for 1975 conditions and run for a 3-year simulation period. Abiotic driving data were derived from data collected on the study site, for the growing season, and from data generated at the nearest U.S. Weather Station, for the remainder of the year. Sulfur dioxide concentrations measured on the study plots (Chapter 3) we used as input to the SO_2 deposition submodel.

The 1976 and 1977 growing seasons were quite different with respect to growing conditions; 1976 was a wet year while 1977 was a dry year. These data were useful to ensure that the model responded properly to extremes of the driving variables. A 3-year simulation was conducted, rather than reinitializing each year, to allow errors that accumulated during 1 simulation year to propagate into the following year. This approach was taken in order to provide confidence that the model could be applied to long-term scenarios.

Model output was compared to field data for aboveground biomass production, sulfur concentration in live leaves, and soil water content in the soil profile. Figures 7.2, 7.3, and 7.4 depict these comparisons for the 1976 and 1977 growing seasons. In most cases model output fell within the statistical variation of the field data (mean ± standard error). The shape and timing of the responses were consistent with those observed in the field.

In addition to these quantiative comparisons, qualitative checks were made on litter dynamics, microbial dynamics, nutrient–pool levels, carbohydrate levels, shoot–nitrogen concentrations, total root biomass, and ruminant body weight. All test results met the criteria established for the objectives of the model.

The primary objective of the SAGE model was to simulate long-term (30 years) dynamics of a mixed-prairie and the impact of exposure to SO_2 on the state and dynamics of the system. Achievement of this objective required that the model respond in a realistic manner to a broad range of driving variables, especially temperature and precipitation, exhibit the proper response to SO_2 over a wide range of concentrations, as well as be capable of producing stable behavior for a period of 30 years. Often, simulation models that produce acceptable behavior for a growing season or a year become unstable when the simulation is continued over a number of years. Errors produced in each simulation year accumulate, producing an imbalance in the system and resulting in unstable behavior. The wide range of tests of the model modules over the broad input ranges necessary for the long-term simulation helped to preclude unstable behavior.

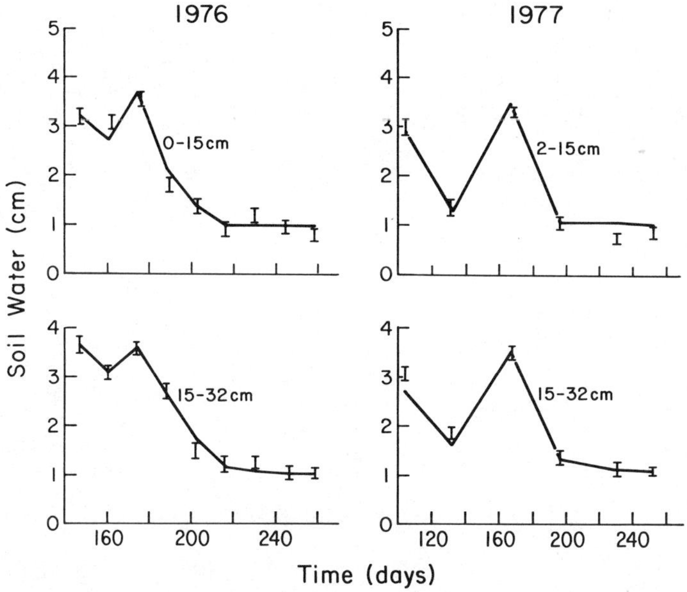

Figure 7.2. Model predictions of soil water content (cm) compared with field data for the 1976 and 1977 growing seasons. Bars represent the mean ± 1 standard error.

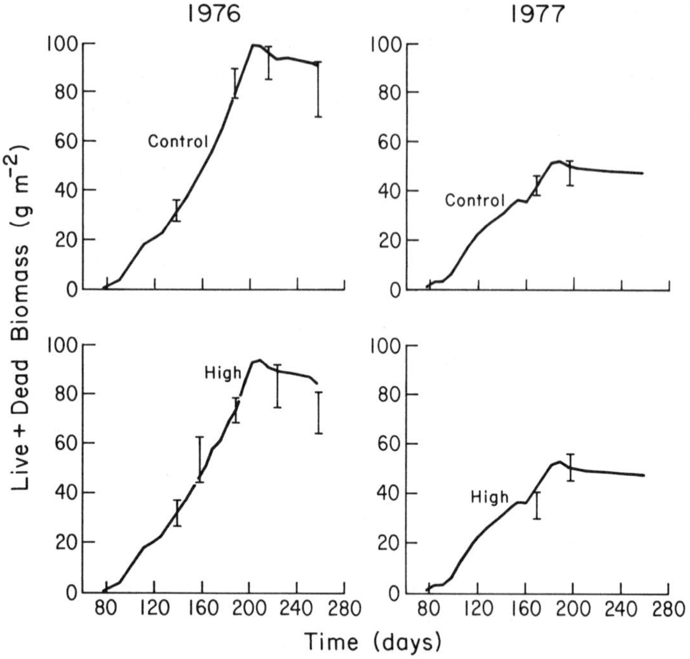

Figure 7.3. Model predictions of live plus recent dead aboveground biomass (g · m^{-2}) compared with field data for the 1976 and 1977 growing seasons. Bars represent the mean ± 1 standard error.

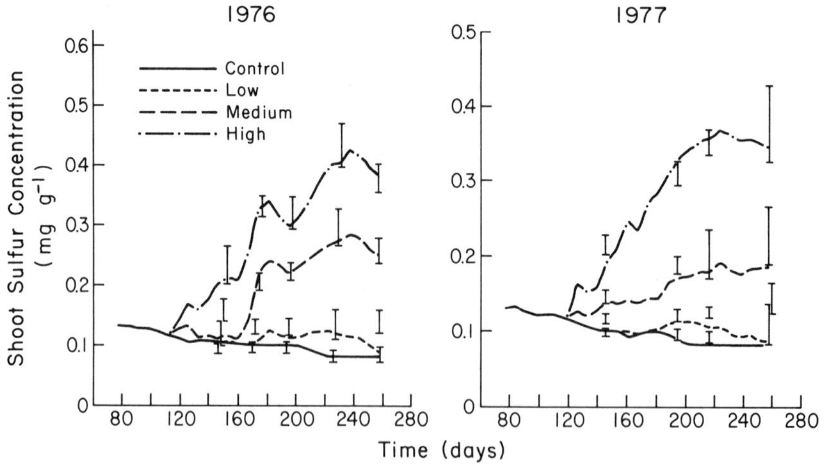

Figure 7.4. Model prediction of shoot sulfur concentrations (mg · g^{-1}) compared with field data for the 1976 and 1977 growing seasons. Bars represent the mean ± 1 standard error.

7.5 Model Experiments

Two kinds of model experiments were conducted. The first began with the model validation procedure already described. The model was initialized for 1975 and run for 3 years through 1977 using experimentally measured values for driving variables and SO_2 concentrations. The model was run for the control and each of the three SO_2 treatments. These model experiments produced as a part of their output, values for variables that were not or could not be measured in the field experiments. Differences between treatments could thereby be explored which were beyond the capabilities of the current field experimentation to detect.

The second kind of model experimentation investigated the impact of long-term exposure of the grassland to SO_2. Two 31-year simulations were conducted to meet this objective. One was a control, with no SO_2 exposure, and the other utilized SO_2 emissions at the federal maximum legal standard (Ludwick et al., 1980). The abiotic driving variables were based on data recorded by the U.S. Weather Bureau (1958–1978) at Billings, Montana. Our objective was to use representative data for the region, rather than duplicate exactly the weather sequence measured over the past 20 years.

Each year was divided into four seasons (fall, winter, spring, and summer) and each season was categorized as wet, dry, or normal based upon precipitation. We then drew samples from these categories to structure a 20-year weather sequence that had the same mix of wet, dry, and normal seasons as was measured at Billings between 1958 and 1978. The conditions for the last 11 years of the simulation were chosen arbitrarily to force stressful weather on the system to test its response to perturbations.

The meterological data used to drive the ecological model were also used to generate 3-hr average SO_2 concentrations through the use of a steady-state Gaussian plume dispersion model (Christiansen, 1975–1976).

7.6 Model Results

7.6.1 Short-Term Impacts

Simulation of the field experimental SO_2 treatments did not indicate any overlooked dramatic or imminently catastrophic effects of the SO_2 exposure. Table 7.1 presents model data for 1976. Proper rules for the display of significant figures have been relaxed in this table as deemed necessary to indicate smaller trends as well as more obvious effects. A consistent decline in gross primary production from 437 to 416 g C · m^{-2} was observed as treatment intensity increased. This difference was less well defined for net primary production because producer respiration declined as well. The lower overall plant activity thereby indicated in response to SO_2 treatment is apparently responsible for the observed very small decline which appeared in the flow of producer carbon to the soil organic matter. A similiarly small decline occurred in all decomposer carbon

Table 7.1. Simulated Impacts of SO_2 Exposure on Carbon, Nitrogen, and Sulfur Flow ($g \cdot m^{-2} \cdot year^{-1}$), 1976

	SO_2 Treatment Level			
Flow Description	Control	Low	Medium	High
Gross carbon				
Atmosphere to producers	437	430	426	416
Producers to atmosphere	309	304	301	291
Net carbon				
Atmosphere to producers	128	126	125	125
Producers to soil organic matter	123.3	122.9	122.7	122.6
Soil organic matter to decomposers	171.4	171.3	171.3	171.2
Decomposers to soil organic matter	48.0	47.9	47.9	47.7
Decomposers to atmosphere	121.0	121.0	121.0	120.9
Nitrogen				
Soil nutrient pool to producers	4.62	4.60	4.60	4.55
Producers to soil organic matter	4.56	4.57	4.58	4.60
Soil organic matter to decomposers	7.70	7.70	7.71	7.70
Decomposers to soil organic matter	3.84	3.85	3.84	3.83
Decomposers to soil nutrient pool	23.17	23.20	23.24	23.27
Soil nutrient pool to decomposers	19.66	19.69	19.72	19.76
Sulfur				
Soil nutrient pool to producers	0.19	0.05	0.05	0.01
Atmosphere to producers	0.05	0.39	0.52	0.72
Atmosphere to soil nutrient pool	<0.01	0.04	0.09	0.13
Producers to soil organic matter	0.42	0.43	0.49	0.56
Soil organic matter to decomposers	0.86	0.86	0.88	0.90
Decomposers to soil organic matter	0.29	0.29	0.29	0.29
Soil nutrient pool to decomposers	9.62	9.65	9.69	9.69
Decomposers to soil nutrient pool	10.20	10.18	10.23	10.25

flows. Flows to and from herbivores were not represented because the 2-year simulation reflected the experiments which did not include grazers.

Nitrogen flows were also affected to a small degree, but sulfur flows to and from producers were substantially changed (Table 7.1). Atmospheric input to producers increased from 0.5 $g\ S \cdot m^{-2}$ for the control to 0.72 $g\ S \cdot m^{-2}$ for the high treatment, while plant uptake from the soil fell from 0.19 to 0.01 $g\ S \cdot m^{-2}$.

The second year of the field experiment simulation showed much reduced producer carbon flows as a result of poorer growing conditions (Table 7.2). Perhaps relatedly, there was a lessened impact on atmosphere-producer carbon flows by SO_2 treatment. However, the reductions in carbon transfers in response to SO_2 exposure from producers to soil organic matter and from soil organic matter to decomposers were larger than those indicated for the first year, though remaining small. Nitrogen uptake from the soil nutrient pool by producers increased in the second year as SO_2 exposure increased, while returns declined; both of these

Table 7.2. Simulated Impacts of SO_2 Exposure on Carbon, Nitrogen, and Sulfur Flow ($g \cdot m^{-2} \cdot year^{-1}$), 1977

	SO_2 Treatment Level			
Flow Description	Control	Low	Medium	High
Gross carbon				
Atmosphere to producers	285	285	285	282
Producers to atmosphere	194	194	194	191
Net carbon				
Atmosphere to producers	91	91	91	91
Producers to soil organic matter	124.5	123.6	123.0	122.3
Soil organic matter to decomposers	180.8	179.6	178.7	178.6
Decomposers to soil organic matter	56.5	55.9	55.5	55.4
Decomposers to atmosphere	124.8	124.3	123.9	124.0
Nitrogen				
Soil nutrient pool to producers	4.74	4.78	4.82	4.87
Producers to soil organic matter	4.89	4.88	4.89	4.87
Soil organic matter to decomposers	8.46	8.44	8.43	8.44
Decomposers to soil organic matter	4.61	4.57	4.54	4.54
Decomposers to soil nutrient pool	23.47	23.53	23.53	23.69
Soil nutrient pool to decomposers	19.67	19.69	19.67	19.80
Sulfur				
Soil nutrient pool to producers	0.24	0.02	0.01	0.01
Atmosphere to producers	0.05	0.24	0.29	0.38
Atmosphere to soil nutrient pool	0.01	0.04	0.07	0.13
Producers to soil organic mater	0.32	0.33	0.45	0.58
Soil organic matter to decomposers	0.82	0.83	0.91	1.01
Decomposers to soil organic matter	0.35	0.35	0.35	0.35
Soil nutrient pool to decomposers	9.44	9.53	9.55	9.61
Decomposers to soil nutrient pool	9.93	10.04	10.13	10.28

trends were opposite to those indicated for the first year. Trends in sulfur flow were similar to the first year, though absolute quantities were much reduced.

Whole-system indicators of SO_2 impact, listed in Table 7.3, showed even less response than the dynamic indicators described above. Producer carbon declined slightly, particularly in the better growth year (1976), with increasing SO_2 treatment. No effects on nitrogen were seen and sulfur changed only in those compartments directly receiving sulfur from the atmosphere.

Ratios of primary production to respiration were slightly different between years, but were not different across treatments within years. Carbon inputs and outputs were balanced in 1976 and showed a net loss of carbon in 1977 ($P/R < 1$). Nutrient cycling ratios (inputs/outputs) showed that nitrogen was balanced and sulfur accumulated in proportion to the exposure concentration.

The decrease in litter disappearance, which was considered to be an important result of the field experimentation (Dodd and Lauenroth, 1981), was slight in the

Table 7.3. System Level Indicators of SO_2 Impact on a Grassland—Simulation Results[1]

	SO_2 Treatments—Years							
System Variable	1976 Control	1976 Low	1976 Medium	1976 High	1977 Control	1977 Low	1977 Medium	1977 High
Producer carbon[2]	286	283	282	277	241	241	241	239
Decomposer carbon[2]	108	108	108	108	111	111	111	111
Soil organic carbon[2]	4874	4874	4874	4874	4883	4883	4883	4883
Producer nitrogen[2]	5.9	5.8	5.7	5.6	4.35	4.31	4.36	
Decomposer nitrogen[2]	9.8	9.8	9.8	9.8	10.1	10.1	10.1	10.0
Soil organic nitrogen[2]	554	553	554	554	554	554	554	554
Inorganic nitrogen[2]	4.4	4.4	4.4	4.4	5.2	5.2	5.2	5.2
Producer sulfur[2]	0.42	0.5	0.63	0.67	0.34	0.41	0.49	0.52
Decomposer sulfur[2]	0.87	0.87	0.87	0.87	0.9	0.9	0.9	0.9
Soil organic sulfur[2]	78.18	78.18	78.18	78	78.9	78.9	78.9	78.9
Inorganic sulfur[2]	17.38	17.54	17.57	17.65	17.72	18.08	18.29	18.41
Gross primary production[3]	437	430	426	416	285	285	285	282
Primary producer respiration	309	304	301	291	194	194	194	191
Carbon reserves[3]	65	65	64.9	64.7	45.3	45.3	45.3	45.5
Decomposer respiration[3]	121	121	121	121	125	124	124	124
Decomposer production[3]	171	171	171	171	181	180	179	179
P–R ratio[3]	1.02	1.01	1.01	1.01	0.89	0.90	0.90	0.90
N-Cycling ratio[3]	0.99	0.99	0.99	0.99	1.00	1.00	1.00	1.00
S-Cycling ratio[3]	1.04	1.06	1.07	1.08	1.03	1.06	1.07	1.09

[1]Values in grams per square meter.
[2]Value at time of peak standing crop of carbon.
[3]End of calendar year values.

simulation results. In the field experiments, significantly greater loss rates from 15 April to 15 July were found for the control as compared to the high-treatment plots. When the experiment was repeated in the following, drier growing season, significantly lower disappearance rates were observed in both the medium- and high-treatment plots. The latter part of the season did not show the differential disappearances, presumably because the overall process was less active at this time.

The failure of this effect to appear in the simulations other than the small decrease in primary production seems a result of the relatively small role of litter in the overall carbon budget for the system. Aboveground litter ranged between 50 and 150 $g\ C \cdot m^{-2}$, while total system carbon averaged 5000 $g\ C \cdot m^{-2}$. Therefore, the hypothesis that the effect of the decreased litter disappearance would reduce nutrient cycling rates and consequently reduce primary production (Lauenroth and Heasley, 1980) was not substantiated by the 2-year simulation.

Many of the results from the field experiments could not be addressed through simulation exercises for one of two reasons. The first is that many of the results were used to define ways in which SO_2 impacted these grasslands. In this case lack of confirmation of an effect would mean an error in formulation or computer coding. The second reason is that the model contained only a subset of the potential relevant variables in the system. The most conspicuous example is the invertebrate component. A substantial portion of the field effort was devoted to the responses of above- and belowground invertebrates to the SO_2 treatments. The most rudimentary representation of invertebrate population dynamics is well beyond our current capabilities either as systems analysts or as invertebrate ecologists. Therefore, we have nothing to report from our modeling experiments on this potentially important topic.

7.6.2 Thirty-Year Simulations

In the long-term simulation, the response of the primary producers was qualitatively, as expected. Sulfur dioxide exposure increased the speed of leaf senescence, and shifted age structure toward younger leaves with higher photosynthetic rates. Hence, gross primary production under SO_2 exposure (Figure 7.5) was often higher than the control. During the wetter growing seasons, however, greater SO_2 uptake resulted in lower gross primary production. Sulfur uptake was from five to six times greater during wet years than dry, accounting for this differential effect. As in the 2-year model experiment, root uptake of sulfur was substantially reduced in the presence of atmospheric SO_2.

Respiration was generally highest in plants exposed to SO_2 (Figure 7.6) because of the energy cost required to disperse the added sulfur which had been taken from the atmosphere. This increase in respiration was occasionally more substantial than the gain in gross production from the presence of younger leaves. In that case, net primary production was reduced for the SO_2 environment (Figure 7.7).

The increased respiration rates of primary producers exposed to SO_2 resulted in a 10 to 20% decrease in labile carbon available for storage (Figure 7.8). This

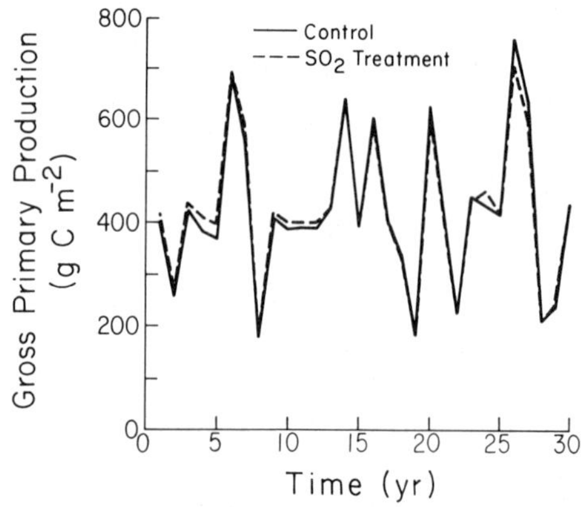

Figure 7.5. Simulated gross primary production (g C · m^{-2}) for a northern mixed prairie exposed to control (no SO_2) and SO_2 (SO_2 concentrations at the Federal Standard level) conditions.

difference was expanding with time of exposure (Figure 7.9). The reduction in carbohydrate storage was expected to increase the susceptibility of the system to stresses such as drought or heavy grazing. Drought was simulated by including 2 years near the end of the simulation with dry springs and summers in sequence. The susceptibility of the system to this stress was measured by the fraction of carbohydrate stores lost during the drought and by the ratio of net primary production during the drought to that of a normal year. In neither case, as a result of this drought, was there more than a 2% difference in susceptibility between SO_2-treated and untreated systems. The mechanisms of response to drought include reduced respiration and the shedding of leaves, hence minimizing the effects of reduced carbohydrate storage during short drought periods. Heavy

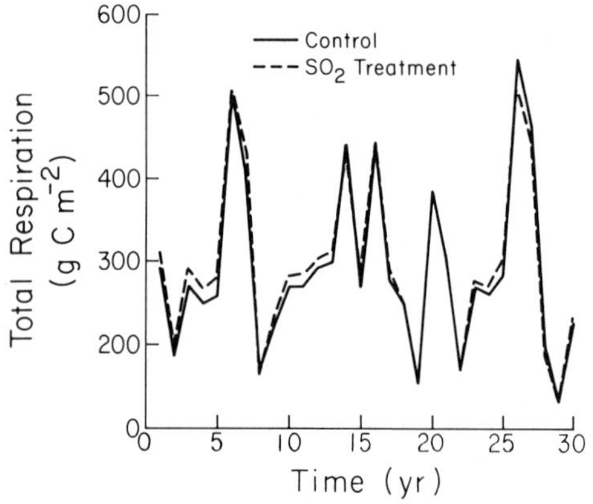

Figure 7.6. Similated total respiration (g C · m^{-2}) for primary producers exposed to control (no SO_2) and SO_2 (SO_2 concentrations at the Federal Standard level) conditions.

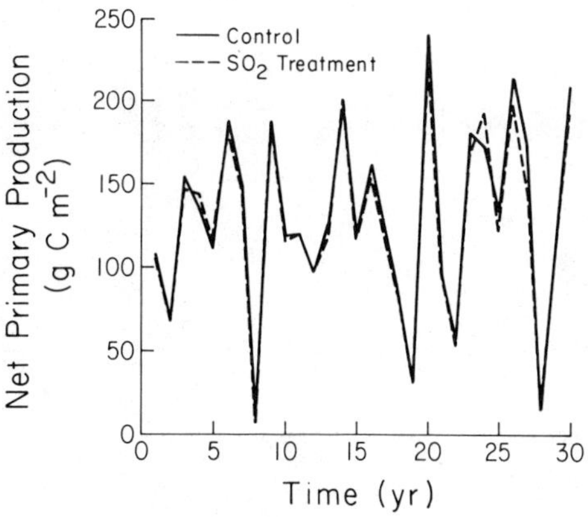

Figure 7.7. Total net primary production (g C · m^{-2}) for a northern mixed prairie exposed to control (no SO_2) and SO_2 (SO_2 concentrations at the Federal Standard level) conditions.

grazing, which would require repeated regrowth during a given season and thereby directly interact with the carbohydrate reserves, still could further increase the primary production differential between the treated and the untreated plots since its stress mechanism is different from that of drought.

Mineralization of inorganic nitrogen was initially 2 to 20% greater in the SO_2-free system than it was for the exposed system. During the first 10 years, nitrogen uptake by the plants exposed to SO_2 was slightly higher than for the control (Figure 7.10), corresponding to the higher demand of the younger structured leaf system. However, as productivity declined for the treated plot, demand for nitrogen declined with it. This resulted in root uptake of mineral nitrogen 7 to 10% less for the SO_2-exposed system than for the control. Hence, the inorganic nitrogen

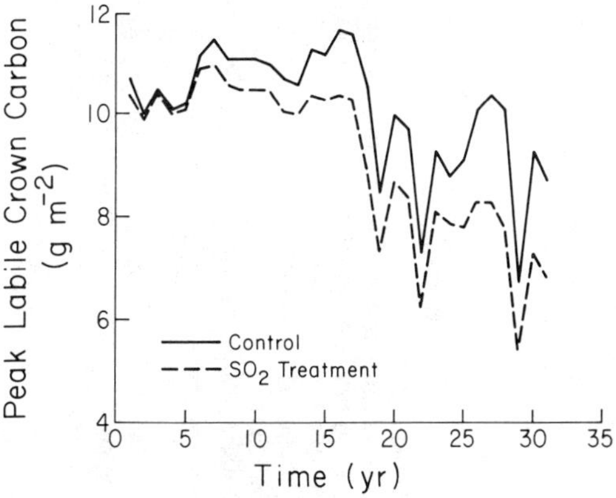

Figure 7.8. Simulated maximum annual labile crown carbon (g · m^{-2}) for a northern mixed prairie exposed to control (no SO_2) and SO_2 (SO_2 concentrations at the Federal Standard level) conditions.

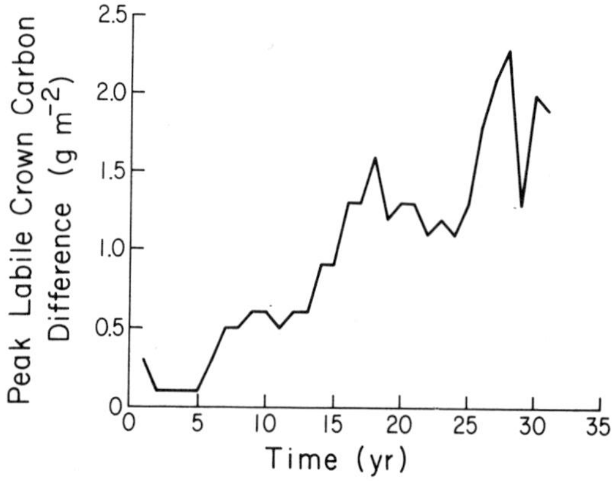

Figure 7.9. The difference in simulated peak labile crown carbon (g · m^{-2}) for a northern mixed prairie exposed to control (no SO_2) and SO_2 (SO_2) concentrations at the Federal Standard Level) conditions.

pool rose to a higher level in the exposed system by the end of the simulation. This rise was aided by the more favorable N:S ratios for microbial mineralization under SO_2 exposure.

Sulfur concentrations increased in the system exposed to SO_2. The increases were differential among system compartments. Sulfur content of the primary producers increased by 10%, but little change occurred for the soil organic matter or decomposers. Over the 30-year period, 15 g · m^{-2} of sulfur was added from SO_2, with almost 90% ending up as inorganic sulfate in the soil. This amounted to an overall 65% increase in the inorganic sulfur pool.

Soil organic matter is recognized as a key indicator of the fertility status and an important determinant of stability of soil structures (Allison, 1965; Voroney et al.,

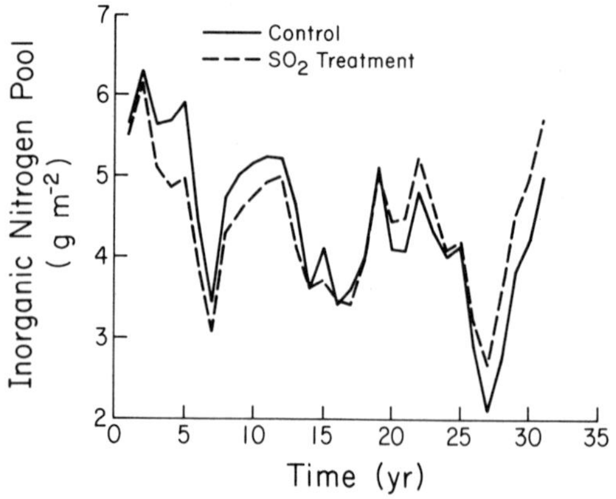

Figure 7.10. Simulated inorganic nitrogen pools (g · m^{-2}) for a northern mixed prairie exposed to control (no SO_2) and SO_2 (SO_2 concentrations at the Federal Standard level) conditions.

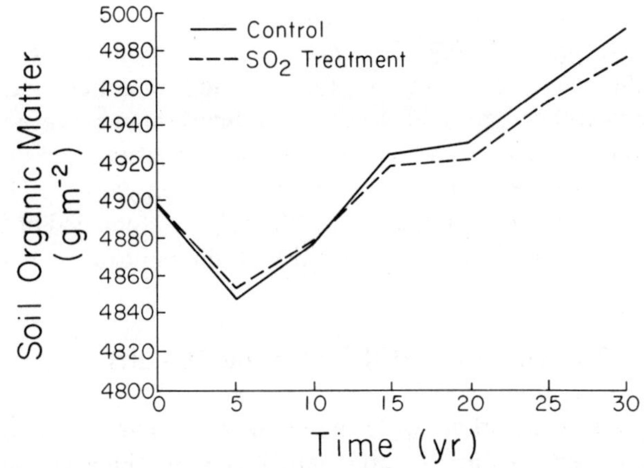

Figure 7.11. Simulated soil organic matter (g C · m^{-2}) for a northern mixed prairie exposed to control (no SO_2) and SO_2 (SO_2 concentrations at the Federal Standard level) conditions.

1981). Changes in soil organic matter over the 30-year simulations showed no significant influence of SO_2 exposure. Soil organic matter (carbon) increased approximately 100 g · m^{-2} for the control simulation, while increasing 75 g · m^{-2} for the SO_2 simulation (Figure 7.11). The difference does not likely imply a significant impact of SO_2, but the magnitude of increase for both exposures is probably an indication of the relative health of both systems.

Cattle production was higher as a result of SO_2 exposure (Figure 7.12). Increasing the sulfur content of the diet of cattle increases their rate of retention of nitrogen as the N:S ratio of the forage approaches that of body proteins. This effect

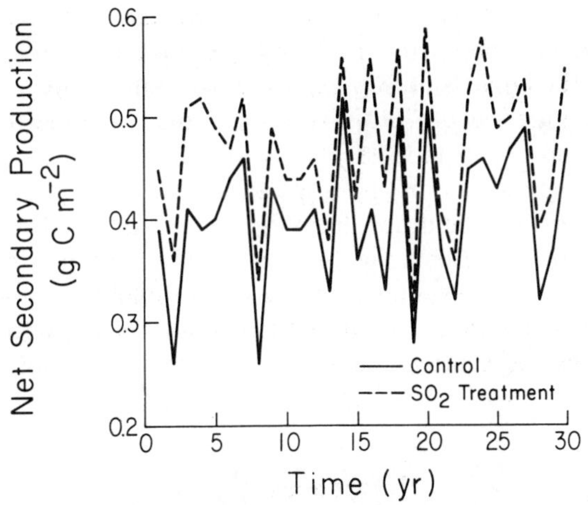

Figure 7.12. Simulated net secondary production (cattle) (g C · m^{-2}) for a northern mixed prairie exposed to control (no SO_2) and SO_2 (SO_2 concentrations at the Federal Standard level) conditions.

has been used to justify the direct supplementation of cattle diets with sulfur (Tisdale, 1977; Garrigus, 1970). However, there was a slight decrease in digestibility for the plants exposed to sulfur in the simulation because of an increase in the senescence rate and a correspondingly higher presence of dead material in the forage. This necessitated an increase in consumption of an average of 3% for the SO_2-exposed system to meet equivalent energy needs. The net result of these effects was a 1 to 14% increase in cattle weights at the end of the growing season for the SO_2-exposed system compared with the control.

7.7 Summary of Simulation Results

The objective of our simulation experiments was to investigate system-level impacts of SO_2 exposure, using a mathematical model which contained representations of SO_2 impacts upon processes. We relied upon the literature and our experimental results for information relating SO_2 exposure to the structure and dynamics of northern mixed prairies. We considered this modeling step to be important to our understanding of impacts of SO_2 on these grasslands because it afforded us the opportunity to evaluate several variables which are either not measurable or not easily measurable in field experiments. It also allowed us to test the long-term (30-year) significance of short-term responses (1–5 years) observed during the field experiment. All of our conclusions from model experiments are limited by the degree to which the model represents the relevant features of the structure and dynamics of the system.

The model experiments produced results with very similar patterns to those observed in the field experiment. Positive, negative, and neutral impacts of SO_2 were mixed. Several of the results from the field experiment that led to speculation about potentially catastrophic impact as they ramified through the entire system and accumulated over time were not supported (Lauenroth and Heasley, 1980). The model behavior over 30 years supported a conclusion based on observations over the 5 years of field experiments; northern mixed prairies have a remarkable capacity to adjust to the impacts of SO_2 exposure on single components. To say it another way, significant effects of SO_2 exposure at the macromolecular (chlorophyll), organ (leaf), and organism (tiller) levels of organization are filtered out as one evaluates their significance at the population, community, and ecosystem levels. A conclusion that can be drawn from this information is that the degree of determinism in the relationships among the behaviors at lower levels of resolution and responses at the upper levels is low. This has major implications for the task of evaluating environmental impacts at the system level, using information about components. One must be cautious about extrapolating from individual experimental results.

References

Allison, F. E. 1965. *Soil Organic Matter and Its Role in Crop Production*. Amsterdam: Elsevier.

Bennett, J. H., and A. C. Hill. 1973. Inhibition of apparent photosynthesis by air pollutants. *J. Environ. Qual.* 2:526–530.
Black, V. S., and M. H. Unsworth. 1980. Stomatal responses to sulfur dioxide and vapor pressure deficit. *J. Exp. Bot.* 31:667–677.
Christiansen, J. H. 1975–1976. Users guide to the Texas episodic model. Data processing section. Texas Air Control Board.
Dodd, J. L., and W. K. Lauenroth. 1981. Effects of low level SO_2 fumigation on decomposition of western wheatgrass litter in a mixed-grass prairie. *Water, Air Soil Pollut.* 15:257–261.
Garrigus, U. S. 1970. The need for sulfur in the diet of ruminants. In *Symp. Sulf. Nutri., 1979*, D. H. Math and J. E. Oldfield, eds., pp. 126–153. Corvallis: Oregon State Univ.
Guderian, R., and H. Van Haut. 1970. Detection of SO_2 effects upon plants. *Staub.* 30:22–35.
Heasley, J. E., W. K. Lauenroth, and J. L. Dodd. 1981. Systems analysis of potential air pollution impacts on grassland ecosystems. In *Energy and Ecological Modelling*, W. J. Mitsch, R. W. Bosserman, and J. M. Klopatck, eds. pp. 347–359. Amsterdam: Elsevier.
Innis, G. S. 1975. The role of total systems models in the grassland biome study. In *Systems Analysis and Simulation in Ecology*, B. C. Patten, ed. Vol. 3. New York: Academic Press.
Innis, G. S. (ed.). 1978. *Grassland Simulation Model.* New York: Springer-Verlag.
Innis, G. S., I. Noy-Meir, M. Godren, and G. M. Van Dyne. 1980. Total-systems simulation models. In *Grasslands, Systems Analysis and Man*, A. I. Breymeyer and G. M. Van Dyne, eds. pp. 759–797. IBP 19 Cambridge: Cambridge Univ. Press.
Kirchner, T. B., and J. M. Vevea. 1983. *PREMOD and MODAID*: Software tools for writing simulation models. In *Analysis of Ecological Systems: State of the Art in Ecological Modelling*, W. K. Lauenroth, G. V. Skogerboe, and M. Flug, eds. pp. 173–183. Amsterdam: Elsevier.
Koong, L. J., and H. L. Lucas. 1973. *A Mathematical Model for the Joint Metabolism of Nitrogen and Energy.* Institute of Statistics Mimeograph Series No. 882. Raleigh, N.C.: North Carolina State University.
Koziol, M. J., and C. E. Jordan. 1978. Changes in carbohydrate levels in red kidney beans (*Phaseolus vulgaris* L.) exposed to sulfur dioxide. *J. Exp. Bot.* 29:1037–1043.
Lauenroth, W. K., and J. E. Heasley. 1980. Impact of atmospheric sulfur deposition on grassland ecosystems. In *Atmospheric Sulfur Deposition—Environmental Impact and Health Effects*, D. S. Shriver, C. R. Richmond, and S. E. Lindberg, eds. pp. 417–430. Ann Arbor, Michigan: Ann Arbor Science Publishers.
Lotka, A. J. 1925. *Elements of Physical Biology.* Baltimore: Williams and Wilkins.
Ludwick, J. D., D. B. Weber, K. B. Olsen, and S. R. Garcia. 1980. Air quality measurements in the coal-fired power plant environment of Colstrip, Montana. *Atmos. Environ.* 14:523–532.
Leetham, J. W., J. L. Dodd, and W. K. Lauenroth. 1983. Effects of low level sulfur dioxide exposure on decomposition of *Agropyron smithii* litter under laboratory conditions. *Water, Air, Soil Pollut.* 19:247–250.
Majernik, O., and T. A. Mansfield. 1971. Effects of SO_2 pollution on stomatal movements in *Vicia faba*. *Phytopathology* 2. 71:123–128.
Margalef, R. 1963. On certain unifying principles in ecology. *Am. Nat.* 97:357–374.
McGill, W. B., H. W. Hunt, R. G. Woodmansee, and J. O. Reuss. 1981. PHOENIX, a model of the dynamics of carbon and nitrogen in grassland soils. *Ecol. Bull.* (*Stockholm*) 33:49–115.
Odum, H. T. 1957. Trophic structure and productivity of Silver Springs, Florida. *Ecol. Monogr.* 27:55–112.
Odum, H. T. 1960. Ecological potential and analogue circuits for the ecosystem. *Am. Sci.* 48:1–8.

Olson, J. S. 1963. Energy storage and the balance of producers and decomposers in ecological systems. *Ecology* 44:322–332.

Parton, W. J. 1978. Abiotic section of ELM. In *Grassland Simulation Model*, G. S. Innis, ed. pp. 31–53. New York: Springer-Verlag.

Patten, B. C. (ed.). 1975. *Systems Analysis and Simulation in Ecology*. Vol. 3. New York: Academic Press.

Patten, B. C., and M. Witkamp. 1967. Systems analysis of ^{134}Cesium kinetics in terrestrial microcosms. *Ecology* 48:813–824.

Ripley, E. A., and R. E. Redmann. 1976. Grassland. In *Vegetation and the Atmosphere, Vol. 2, Case Studies*, J. L. Monteith, ed. pp. 349–398. London: Academic Press.

Schramel, M. 1975. Influence of sulfur dioxide on stomatal aperatures and diffusive resistance of leaves in various species of cultivated plants under optimum soil moisture and drought conditions. *Bull. Acad. Polomaise Sci. Ser. Sci. Biol.* 23:57–66.

Swift, D. M. 1983. A simulation model of energy and nitrogen balance for free-ranging ruminants. *J. Wildl. Manage.* 47:620–645.

Tisdale, T. S. 1977. *Sulfur in Forage Quality and Ruminant Nutrition*. Tech. Bull. 22. Washington, D.C.: The Sulfur Institute.

Unsworth, M. H., P. V. Biscoe, and H. R. Pinckney. 1972. Stomatal responses to sulfur dioxide. *Nature (London)* 238:458–459.

U.S. Weather Bureau. 1958–1978. *U.S. Climatological Survey*. Billings, Montana.

Van Dyne, G. M. 1966. *Ecosystems, systems ecology and systems ecologists*. Oak Ridge, TN: ORNL/3957.

Van Dyne, G. M. (ed.). 1969. *The Ecosystem Concept of Natural Resource Management*. New York: Academic Press.

Van Dyne, G. M. 1972. Organization and management of an integrated ecological research program with special emphasis on systems analysis, universities and scientific cooperation. In *Mathematical Models in Ecology*, J. N. R. Jeffers, ed. pp. 111–172. Oxford: Blackwell.

Van Dyne, G. M. 1978. Foreword: Perspectives on the ELM model and modeling efforts. In G. S. Innis, ed. *Grassland Simulation Model*. Ecol. Studies 26. New York: Springer Verlag.

Voroney, R. P., J. A. van Veen, and E. A. Paul. 1981. Organic C dynamics in grassland soils. 2. Model validation and simultion of the long-term effects of cultivation and rainfall erosion. *Can. J. Soil Sci.* 61:211–224.

Watt, K. E. F. 1966. *Systems Analysis in Ecology*. New York: Academic Press.

White, K. L., A. C. Hill, and J. E. Bennett. 1974. Synergistic inhibition of apparent photosynthesis rate of alfalfa by combinations of sulfur dioxide and nitrogen dioxide. *Environ. Sci. Tech.* 8(6):574–575.

Ziegler, I. 1975. The effect of SO_2 pollution on plant metabolism. *Residue Rev.* 56:79–105.

8. Sulfur Dioxide and Grasslands: A Synthesis

W. K. Lauenroth, D. G. Milchunas, and T. P. Yorks

8.1 Introduction

We began this volume with the assertion that grasslands have a large capacity to adjust to perturbations. Evidence to support this can be found from experimental results as well as from the general observation that humankind has been more successful managing grasslands than other ecological systems. Indeed, grasslands are the foundation of modern agriculture.

In this chapter we evaluate several important characteristics of grasslands that may partially explain their ability to adjust to perturbations. We begin by summarizing the significant findings from our field, laboratory and simulation modeling experiments. We then interpret these findings in relation to energy development in the northern Great Plains of the United States. We end with a discussion of the significance of our results for other grasslands and make comparisons with other vegetation types.

8.2 Summary of Findings

8.2.1 Sulfur Deposition

While SO_2 concentrations are a useful comparative variable for assessing impacts, the deposition and distribution of the atmospheric sulfur are more relevant because they can vary between and within communities for a given atmospheric concentration, and they are more directly related to component and system responses. The concentration and accumulation of sulfur by the vegetation increased with increasing level and duration of exposure, and were further influenced by the productivity during a particular growing season, the nutrient status of the soil, and defoliation treatments. Increases in plant sulfur content cannot necessarily be related to detrimental SO_2 impacts. The increase in sulfur concentration of plants on the low- and medium-SO_2 treatments was accompanied by an increase in organic sulfur concentration (Figure 4.9). It was only on the high-SO_2 treatment that organic sulfur concentrations decreased, indicating direct toxic effects of exposure.

Translocation of sulfur belowground was affected by SO_2 exposure late in the season after shoots had accumulated large quantities of sulfur. At this time, less sulfur was translocated belowground on the high-SO_2 treatment, yet belowground total sulfur remained unchanged and organic sulfur decreased (Section 4.3.4). The functional equilibrium between roots and shoots was upset, resulting in a disruption of the normal nutrient supply relationship between organs that cannot function independently.

Evidence indicated that the nutrient and carbon transfer processes between system components, as well as within autotrophic components, were disrupted by high-SO_2 concentrations. Elevated sulfur content of surface litter (Figure 4.13) and depressed soil pH (Table 4.11) reduced the rate of organic matter decomposition. The reductions in decomposition and soil pH occurred after 5 years of exposure on the high-SO_2 treatment, where deposition of SO_2–sulfur to the system was estimated to be 0.58 g S $\cdot$ m^{-2} $\cdot$ $year^{-1}$ (Table 4.2). Deposition of sulfur attributable to SO_2 treatment was only 0.13 g S $\cdot$ m^{-2} $\cdot$ $year^{-1}$ on the low-SO_2 treatment. Three quarters of the SO_2–sulfur deposition during the growing season was a result of uptake by the vegetation. This sulfur uptake varied as a function of the annual productivity. Sulfur dioxide uptake was greatest in those years with the most favorable growing-season weather.

8.2.2 Vegetation

The northern mixed-prairie vegetation was impacted in a variety of ways by exposure to SO_2. Several broad generalities can be made about these responses. We were much more successful in measuring statistically significant impacts of SO_2 at the lowest levels of organization than at the highest. A large number of the responses that we measured showed stimulation at the lowest SO_2 concentration and either no departure from the control at the highest concentration or a negative response. Several responses changed from no effect to stimulation and finally to a negative effect as the duration of exposure increased.

Sulfur dioxide exposure resulted in significant plasmolysis and reductions in the cover of lichens (Section 5.2.1.3); a mixture of significant increases and decreases in chlorophyll concentration in vascular plants (Figures 5.3 and 5.4); stimulated carbon translocation aboveground and to belowground organs (Tables 5.4 and

5.5); accumulation of large quantities of sulfur in aboveground organs (Figure 4.9); increased leaf areas (Figure 5.16), and numbers of leaves (Figure 5.22); and no measurable impacts on species composition with the exception of *B. japonicus* (Table 5.2), or productivity (Table 5.6).

Several of these responses represent evidence that the plants have a large capacity to compensate for small deleterious impacts of SO_2. *Agropyron smithii* showed decreased chlorophyll content and increased rates of leaf senescence as a result of SO_2 exposure. The compensation for the impact appeared as a larger number of leaves on each tiller and a greater live leaf area per tiller. A series of such apparently compensatory relationships can be found in our results.

An example of the influence of level of organization on the significance of SO_2 effects can be demonstrated with chlorophyll. The relationship between chlorophyll and photosynthesis is direct in a biochemical sense. The relationship between chlorophyll concentration in the leaves of plants and photosynthetic rate and/or biomass production is less direct. Decreases in the content of chlorophylls *a* and *b* in leaves of *A. smithii* (Figure 5.3) were not translated into decreases in biomass production (Table 5.6). Additionally, chlorophyll concentrations showed gradients of responses to SO_2 exposure related to the treatment concentration gradient and the time of exposure. Early in the growing season, chlorophylls *a* and *b* were increased by exposure to the lowest SO_2 concentration and decreased by the high concentration. As the season progressed differences among SO_2 treatments disappeared and all were lower than the control.

8.2.3 Heterotrophs

Total arthropod populations were not significantly affected by SO_2 exposure, but some individual species and/or groups were affected. The more important principles that emerged are the particular sensitivity of organisms intimately associated with soil water, the differential sensitivity of organisms found near the surface relative to those more abundant in the deep soil layers, and the seasonal effects of SO_2 exposure in relation to activity patterns of the organisms.

The groups of soil water-associated organisms that we examined were in some way affected by SO_2 exposure. Tardigrades and non-stylet-bearing nematodes showed significant population declines in the upper (0- to 10-cm) soil layers, but no perceptible effects below 10 cm. Population increases were noted for the stylet-bearing nematodes leaving the overall population apparently unaffected. Rotifers responded positively to low concentrations, and negatively to high concentrations, of SO_2.

Responses of taxa comprising either soil microarthropod or aboveground arthropod groups were both positive and negative, resulting in no net effect of SO_2 on either group as a whole. For the soil microarthropods that were negatively impacted by SO_2, the effects were most apparent when high-soil water levels supported large, active populations.

For the closely studied grasshoppers (*Melanoplus and Eritettix*), laboratory observations suggested that physiologically marginal individuals could be selectively affected by SO_2 exposure and thereby eliminated. Additionally, palatability of leaves exposed to SO_2 was reduced, and avoidance was a possible reason for decreased egg-laying activities on more heavily treated experimental plots.

At the community level, plant feeding types of soil microarthropods increased and aboveground plant sap feeding arthropods declined, possibly in response to the differential accumulation of sulfur in above- and belowground plant organs. The effects of SO_2 exposure on decomposition of plant material may have been responsible for the observed decline in the fungivores. Aboveground parasite–parasitoid arthropods were negatively affected as SO_2 exposure level increased. The observed decline in the pollen–nectar feeding group of arthropods may have occurred because of a special sensitivity of these organisms to SO_2.

8.2.4 Simulation Modeling

Modeling exercises were conducted for two sets of conditions. The first simulated the field experiments using initial conditions, SO_2 concentrations and meteorological data from the years of the experiments. These simulations were carried out for validation purposes (Figures 7.2, 7.3, 7.4) and also to obtain values for variables that were not measured in the field (Tables 7.1 and 7.2). Experiments of the second kind were conducted to explore system responses over long periods of SO_2 exposure (30 years) (Figures 7.5–7.12). SO_2 concentrations were generated utilizing a gaussian plume dispersion model. A time series of SO_2 concentrations was generated under the constraint that the federal secondary standard was not violated. This resulted in an annual average concentration of 78 $\mu g \cdot m^{-3}$.

Neither modeling exercise produced any surprising results. Simulations of the field experiment resulted in small decreases in plant carbon and nitrogen and substantial increases in sulfur. Decomposer carbon, nitrogen, and sulfur were uninfluenced by SO_2 exposure. Soil carbon and nitrogen were unchanged and inorganic sulfur increased slightly. Both gross primary production and plant respiration were depressed slightly by SO_2 exposure. The result was that net primary production was unchanged. Carbon reserves of plants and decomposer respiration and production were very slightly decreased by SO_2 exposure. To a large extent these simulations reinforced the conclusions reached through field experiments.

Our interest in long-term simulations was centered around the anticipation that many of the impacts of SO_2 exposure documented in the field experiment would have substantial influence on system structure and function if allowed to accumulate through time.

Gross and net primary production as well as plant respiration behaved very predictably over the 30 simulation years. Small interactions with specific weather years occurred but never with the effect of creating a divergence in the trajectories of the two conditions (with or without SO_2). Net secondary production, which for this model was cattle production, showed substantial differences between the control and SO_2 simulations especially in certain weather years. We did not have data to validate this response from the field experiment. The explanation resides in the beneficial effect of increased sulfur content of forage on nitrogen retention by ruminants.

Positive impacts were also found for soil organic nitrogen. Soil inorganic nitrogen was decreased slightly during the first 10–15 years of simulation (Figure 7.10). The increase after 30 years likely represents a new steady state resulting from the decrease in aboveground decomposition caused by SO_2.

The most important negative impact of SO_2 was its influence on carbon reserves stored in the crowns of perennial grasses. We measured a decrease in rhizome biomass of *A. smithii* (Table 5.2). This was one of the only responses from the field experiment that seemed to accumulate through time. This represents an impact that was not neutralized by the system rapidly seeking a new steady state. After 30 years the difference between the SO_2 simulation and the control was 2 g $C \cdot m^{-2}$ or almost 20% of the control value. The consequences of continuation of that trend would be either local elimination of the present perennial grasses or more likely, selection for those components of the population which were most resistant to SO_2 exposure.

8.3 Conclusions

We can draw two kinds of conclusions from this work. The first relates to inferences about the field experiments. These are relatively straightforward and relate to the short- and long-term consequences of exposing northern mixed prairies to sulfur dioxide concentrations similar to our experimental concentrations. The second set of conclusions relates to a broader topic: what are the likely consequences of exposing northern mixed prairies to sulfur dioxide concentrations resulting from the current level of energy development in the region and to concentrations projected for future levels of energy development? This latter task involves both interpolation and extrapolation of our experimental and simulation modeling results.

One of the most difficult tasks in generalizing our findings to energy development scenarios is to find a common basis for expressing SO_2 exposure. Our experimental treatments resulted in nearly continuous exposure of the plots to SO_2 for approximately 7 months each year for 15 years. Concentrations varied with weather conditions, particularly temperature and wind, but we still had a much lower frequency of zero SO_2 concentrations than one might expect for any location in the northern Great Plains even under maximum development scenarios. The wind data that we used as input to a Gaussian-plume diffusion model to generate time series of SO_2 data for modeling exercises indicated that, at 25 km from the stack, the site that was at greatest hazard was exposed to zero SO_2 for 83% of the 3-hr periods in each year. Peak concentrations for our simulations (78 $\mu g \cdot m^{-3}$ ann. avg.) were 1150 $\mu g \cdot m^{-3}$ occurring during 0.06% of 3-hr periods. The highest 3-hr average for our most comparable treatment (low) ranged from 460 to 1615 $\mu g \cdot m^{-3}$ (Table 3.1). Peaks in 4 of the 5 years were less than 1000 $\mu g \cdot m^{-3}$. Peaks for the medium and high treatments ranged up to 6800 $\mu g \cdot m^{-3}$.

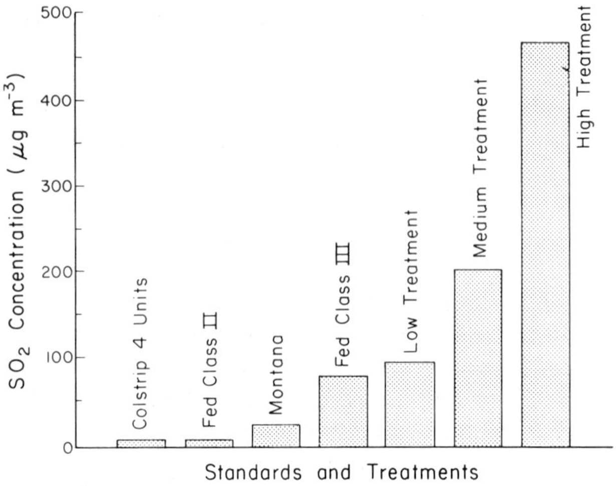

Figure 8.1. Comparison among annual arithmetic mean SO_2 concentrations ($\mu g \cdot m^{-3}$) of various federal and state standards, Colstrip (Montana) power plant (Ludwick et al., 1980) and the SO_2 treatments.

Annual arithmetic mean concentrations provide a method of comparing exposure regimes. The data in Figure 8.1 were calculated assuming our treatments were continued for the entire year. This is a reasonable assumption for direct and many of the indirect impacts upon the biota. This method distorts the likely effects on soils, especially pH. Direct deposition of SO_2–sulfur to the soil was approximately 22% of the growing-season total, and we would expect this to be greater with diminished winter canopy cover. Despite this limitation we believe that annual averages have value for comparisons.

Our interpretation of the relationship of our experimental treatments to standards and development scenarios is that the low concentration treatment is an upper bound for the federal standard assuming occasional violations. The medium- and high-concentration treatments may be useful to explore the potential consequences of relaxing federal or state standards. All of our treatments represent annual average SO_2 concentrations one to two orders of magnitude greater than the standards that currently apply to the greatest portion of the northern Great Plains. The Class II PSD increment is 5.7 $\mu g \cdot m^{-3}$ annual average, a maximum 1-hr peak of 38 $\mu g \cdot m^{-3}$ and a maximum 3-hr peak of 267 $\mu g \cdot m^{-3}$. The Montana State Standard, 20 $\mu g \cdot m^{-3}$, is only 20% of the low-treatment SO_2 concentration.

8.3.1 Field Experiments

Exposure of northern mixed-prairies to concentrations similar to our high concentration treatment for a long period of time (>5 years) would very likely produce large negative impacts. The trends after 5 years of SO_2 exposure, in such variables as soil surface pH, mesofauna populations, sulfur concentrations in plants and litter, aboveground decomposition, and carbon storage in rhizomes of perennial grasses, if continued may result in drastic changes in system structure and behavior. We cannot say with certainty that system collapse would be imminent, but we would expect substantial alterations. The capacity of the system to adjust to this impact is not certain for a long period of time. In our 5 years of experimentation we did not record fundamental shifts in behavior at this SO_2 concentration. We did make a number of observations, listed above, which indicated that the capacitv of the system to adjust to SO_2 impacts was being exceeded.

The medium SO_2 concentration produced less clear results than the high-concentration treatment. It was clear from several of our experiments that the biota was sensitive to concentrations within this range. Particularly sensitive were leaf chlorophyll, soil surface pH, plant and litter sulfur concentrations, and aboveground decomposition. Conclusions about this treatment are difficult to make because for many of our experimental response variables we concentrated upon the low and high treatments.

The low-SO_2 concentration elicited primary producer responses which, over the short-term, can be described as stimulation from nutrient enrichment. Optimal nitrogen to sulfur ratios (Table 4.6) and increased organic sulfur synthesis (Figure 4.9) were nutritional characteristics which could be linked to increased canopy height and fullness (Figure 5.1) and leaf growth (Figure 5.17). Differences between the low-SO_2 treatment and the control could not be distinguished for many of the other variables we measured. Exceptions were, decreases in lichen cover (Figure 5.7) and biomass of *Bromus japonicus* (Figure 5.24), and indications of lower soil pH (Table 4.11) and litter decomposition (Table 4.9). The small increases in soil sulfate concentrations on the low- and medium-SO_2 treatments compared to the control and the very large increase in the high-SO_2 treatment suggested that a conversion and/or immobilization mechanism was in operation which became saturated only at the high-treatment levels. The slight decrease in soil pH on the low-SO_2 treatment probably represents a lag between sulfur deposition and subsequent incorporation into organic constituents or adsorption onto mineral colloids.

The fact that ecological systems can adjust to perturbations is a potential complicating factor in assessing air pollution effects. Our results indicate that impacts of SO_2 upon individual components of a system may be filtered out as one examines increasingly aggregated responses. Decreases in both chlorophyll content and longevity of leaf blades did not result in decreased biomass production by *Agropyron smithii*. Decreased litter decomposition did not decrease either nutrient availability or net primary production.

In some cases one could argue that the difference was related to our ability to measure the various responses. Perhaps chlorophyll content can be measured more precisely than aboveground biomass. In other cases it seems clear that the system was sufficiently flexible to adjust to the perturbation. This raises questions about the value of assuming simple serial causal chains connecting the responses of components to the response of the system.

The implication of this for environmental impact assessment is that extrapolation from small scale effects to large scale responses will seldom be easy. Biological indicators of air pollution effects should be selected carefully to insure that they will provide information at the scale of interest.

8.3.2 Consequences of Energy Development

Generalization of our experimental results and the current air pollution literature into a framework which will allow us to make conclusions about the probable impacts of energy development on northern mixed prairies require a clear statement of assumptions concerning the realtionship between exposure concentration and exposure duration. With respect to SO_2 concentration, we recognize two categories that share a fuzzy boundary. The first is high concentrations. In this case SO_2 is delivered to the system at a rate exceeding the capacity to convert it to nontoxic forms. The second is low concentrations. Here the delivery rate does not exceed the conversion capacity. Exposure durations are also separated into two classes: short-term; those which have the characteristic of a single episode and, long-term; those which have the characteristics of a time series of episodes or a relatively constant exposure.

Much of the early work on the effects of SO_2 on ecological systems or components was focused upon responses to high concentrations for short time periods. These experiments were concerned with understanding the short-term events which were common near power plants or smelters with relatively short stacks or in heavily industrialized areas. For a number of years these were considered to be the only conditions under which SO_2 adversely affected ecological systems. When we began this study in the early 1970s, controversy remained regarding the existence of "invisible injury" or adverse affects on plants at relatively low concentrations of SO_2. Currently there is no longer a question of whether plants or ecological systems respond to low concentrations of SO_2. The important questions concern the significance of these responses. With current regulations of emission, the importance of the so-called "acute exposures" is considerably diminished. Our conclusions assume that an event with concentrations approaching 1300 $\mu g \cdot m^{-3}$ for an hour or more would be an extremely low probability event for the northern Great Plains even under the maximum development scenarios.

The condition that we believe will be most relevant for future energy development and ecological impacts in the northern Great Plains will fall into the categories of low-exposure concentrations (well below 1300 $\mu g \cdot m^{-3}$) and long durations. These durations will have the characteristics of a time series of episodes rather than a constant exposure. Under these conditions there are several responses we might expect to observe. The first is no detectable change in system behavior or structure. The interpretation here is that either SO_2 concentrations are below the detection limit of the biota or responses are sufficiently small that they are obscured by normal variability. A second category of responses to be expected is the positive or subsidy response. In this case the increased availability of sulfur to the biota results in enhancement of system activity. The final category of response is the negative impact. This negative effect is not the direct result of SO_2 but rather occurs through the accumulation of SO_2 derivatives that become toxic only when they are present in high concentrations.

These three potential responses to low concentrations of SO_2 are not exclusive categories. A single system component or process may progress from no observable impact to a positive response and finally a negative response as the time of exposure increases. This phenomenon may occur within a single growing season or it may take many growing seasons. We expect this kind of response when the pollutant is also an important energy resource or nutrient for the system. Odum et al. (1979) described this as a subsidy stress gradient. Their hypothetical performance curve was a parabola opening downward with the maximum coinciding with the peak of the subsidy or in the case of SO_2, the maximum nutrient effect.

We consider that both the positive and the negative responses to SO_2 represent perturbations to ecological systems. The responses that we interpret as positive can represent similarly significant challenges to the system as the negative impacts. Both situations create conditions under which nominal relationships among system components must be adjusted to incorporate the new responses. As we stated earlier, grasslands have been shown in the past to have a large capacity to adjust to environmental fluctuations. We expect that they will prove resilient to both positive and negative responses to air pollution impacts.

Our results from the low concentration treatment showed all three responses as well as a high degree of resilience with respect to SO_2 impacts. For time periods equal to the duration of our experiment, we found no evidence that would lead us to the conclusion that adverse system responses can be expected from SO_2 exposure concentrations lower than our low treatment. For time periods longer than our experiment and up to the length of our simulation experiments (30 years) we are less certain but we still have no reason to predict adverse responses. We would expect a mixture of positive and negative responses of system components, but we believe that these grasslands are sufficiently resilient to adjust to these changes. A caution that must be raised here is that while grasslands are the most extensive vegetation type in the northern Great Plains, other types are present (see Figure 2.4, Section 2.4.1) and may be more sensitive to SO_2 exposure. This is particularly true for the pine forests that occupy the ridge tops over much of the area (see Section 8.5).

We can state these conclusions in a stronger fashion if we assume that current air quality criteria will remain intact, including PSD increments. Under these conditions we would predict that long-term exposure of northern mixed prairies to SO_2 from energy development will very likely have no measurable negative impacts.

Holling (1978) cautions that since everything is not indiscriminately connected to everything else in ecological systems it is possible to reach a conclusion of no negative impacts in an environmental assessment because the wrong variables were measured. Our approach emphasized structural and functional features of the grasslands rather than life histories of indiviudal species. We considered carbon, nitrogen, and sulfur cycling to be important processes which rely upon many of the connections among system components that we considered to be important. We readily admit, however, that we may not have evaluated the "right" variable, the one which would alert us to important or imminent drastic changes in the grassland system.

Another limitation of our conclusions is that we did not consider pollutants other than SO_2. Some evidence is accumulating which indicates that interactions among pollutants may produce greater negative impacts upon ecological systems than single pollutants (Reinert et al., 1975). Projections by Durran et al. (1979) anticipated a severalfold increase in sulfur oxides between 1975 and 1985 and significant increases in nitrogen oxides and hydrocarbons. They indicated concern for air quality impacts associated mostly with sulfur oxides. It is not possible at this time for us to predict the consequences for grasslands of interactions among pollutants in the northern Great Plains.

8.4 Comparisons with Other Ecological Systems

Ecological systems differ in their sensitivity to SO_2 exposure depending upon a variety of structural and dynamic features. In this section we will discuss our results in terms of their significance for grasslands and other ecological systems. Because we and others emphasized responses of plant communities, we will focus on the autotrophs.

8.4.1 Grasslands

Grasses, grasslands, and the herbivores that coevolved with them represent the major food resource for the world (Singh et al., 1983a). Grasslands on each continent are usually found between deserts and forests. The seasonal distribution of soil water availability is the best single predictor of the presence or absence of grasslands. They occur in regions which have a season during which soil water falls below the requirements for forests yet they receive sufficient precipitation to sustain grasses (Lauenroth, 1979).

Worldwide, grasslands potentially occur on 30.6 million km^2 (FAO 1979). In North America they are the potential natural vegetation on 3.74 million km^2. Based upon the distribution of climates, semiarid grasslands represent half of the global total and slightly more than half of North American grasslands (Bailey, 1979). Although figures are not available for actual area occupied by grasslands, we would expect the semiarid grasslands to be a larger proportion of remaining grasslands than they are of the potential grasslands. This is because of the heavy exploitation of subhumid grasslands for crop production. In North America, the corn belt is located in a region of previously subhumid grasslands.

The northern mixed prairie of the Great Plains of North America is a temperate-semiarid grassland type (Singh et al., 1983b). Dominant genera of plants include *Agropyron, Koeleria, Stipa, Poa, Taraxacum,* and *Artemisia*. Growth and biomass accumulation are most rapid during the coolest portions of the growing season. This growth pattern is consistent with characteristics of the C_3 photosynthetic pathway, which the dominant species possess, and also with the seasonal pattern of precipitation.

All of the temperate-semiarid grasslands in North America share several to all of these characteristics with the northern mixed prairie. In addition, North American grasslands have considerable overlap in vertebrate and invertebrate fauna. We would expect a high degree of similarity in responses to SO_2 exposure among temperate North American grasslands. Alternatively, we would not be surprised to find differences in response as a result of site- and species-specific characteristics. We have no reason to believe that grasslands at any temperate-semiarid location in North America should be substantially more sensitive to SO_2 exposure than the northern mixed prairie.

Globally, temperate-semiarid grasslands are found on each continent except Africa and Europe (Singh et al., 1983a). In South America they are found at high elevation in the Andes and associated with the Patagonian–Fuegian–Falkland steppe. The latter region contains species of *Agropyron, Stipa, Poa*, and *Koeleria*. Australia contains dry temperate grasslands with species of *Stipa* and *Poa*. Temperate-semiarid grasslands encompass a substantial portion of central Asia, extending from China to the Black Sea. These grasslands are very similar in many characteristics to the northern mixed prairie. Dominant genera throughout the region include *Agropyron, Stipa, Poa, Koeleria*, and *Artemisia*.

We predict that responses of temperate-semiarid grasslands worldwide should be quite similar to those we observed in our field experiment. We base that speculation upon the high degree of structural and behavioral similarity that has been reported for this group of grasslands. We will address this point in greater detail in the next section.

8.4.2 Forests

Ecological systems with significant components of lichens and evergreen coniferous trees appear to be the most sensitive to air pollutants (Miller and McBride, 1975; Linzon, 1978; Smith, 1981). Effects on the distribution and abundance of lichens around industrial and urban complexes were among the earliest records of air pollution impacts upon plants and systems (Linzon, 1978). Lichen "deserts" are often described around air pollution sources. Our results also show lichens as clearly among the most sensitive plants (Figure 5.7) to SO_2 exposure.

Evergreen conifers, perhaps because of the longevity of their leaves, are quite sensitive to cumulative effects of SO_2 air pollution. A commonly recorded response of pines to SO_2 exposure is decreased needle retention (Kozlowski, 1980). Miller and McBride (1975) and Smith (1981) summarized a number of documented cases of significant impacts of air pollution on forests. A large number of these were associated with SO_2 exposure. A result of exposure to very high concentrations of SO_2 for several to many years was a series of elliptical zones of impact around the source in a downwind direction. The innermost zone was characterized by a lack of vegetation and significant soil erosion. Surrounding this was a zone of very resistant plants most often containing no lichens or trees. Next occurred a zone of resistant trees followed by the expected normal forest vegetation. Often the zone with very resistant plants is dominated by grasses (Guderian and Kueppers, 1980; Scale, 1980; Smith, 1981). The presence of grasses under these conditions is probably related to their stature and relative drought resistance. The low height of grasses brings them in contact with a smaller cross section of the polluted atmosphere, and their relative drought resistance very likely results in morphological and physiological adaptions which reduce pollutant impacts.

An important characteristic that grasses share with other suffrutescent life forms is high intra- and interseasonal turnover of aboveground organs. In contrast to conifers, the expected functional life of a grass blade is often a fraction of a single growing season (Figure 2.8). Injury to or even complete loss of a blade as a result of SO_2 exposure is likely to have a small negative effect on growing-season carbon balance. Adaptations for regrowth following grazing may be important in responses following SO_2 injury to a leaf. A similar loss to a conifer with its highly deterministic leaf initiation capabilities will represent a much more significant impact upon annual carbon balance. Additionally, in temperate regions, grasses are photosynthetically active for only a portion of the year while the most sensitive conifers retain green leaves and the potential for gas exchange year-round. Scheffer and Hedgcock (1955) reported that the deciduous conifer *Larix occidentalis* had greater survival capabilities in an SO_2-polluted environment because foliage was not present in the winter.

At the outset of this chapter we asserted that grasslands were characterized by a large capacity to adjust to perturbations. We will not return to that subject to explore the source of that resilience. In a review of impacts of air pollutants upon forested ecosystems, Kozlowski (1980) stated "Change in structure of a relatively stable forest because of low-level pollution tends to be subtle and gradual at first. However, the structure of delicately balanced forest ecosystems often depends on a few critical species." We have a quite different interpretation of grasslands. In general, grasslands do not seem to be profitably analyzed by such an equilibrium-centered stability approach. We do not conceptualize northern mixed prairies as poised upon the brink of disaster with respect to environmental perturbations. We prefer to employ a multiequilibria model (Margalef, 1975; Holling, 1978). This model depends upon low stability of any single equilibrium state and high mobility among states. The ability of the system to move among a number of similar states in response to perturbations is its resilience.

An important factor in the persistence of any ecological system is related to its ability to attenuate environmental variability. The processes that achieve this in most ecosytems are related to a large relatively inert reservoir of organic matter (O'Neill and Reichle, 1980). In forests this consists of large quantities aboveground in standing dead wood (tree trunks), litter and soil organic matter. Because of the structure of forests, a large portion of the organic matter reservoir remains aboveground. By contrast the organic matter stores of grasslands are concentrated belowground. Ajtay et al. (1979) estimated soil organic carbon content (to a depth of 1 m) for temperate forests to be 12,000 g · m^{-2} and for temperate semiarid grasslands to be 30,000 g · m^{-2}. This is considerably greater than the estimate we used in our simulation model (Table 7.3) and we have reason to believe now that our initial estimates were low (Van Veen and Paul, 1981). The amounts of aboveground litter in forests and grasslands shows the opposite trend. Temperate forests are estimated to contain an average of 3000 g · m^{-2} of aboveground litter while grasslands contain 300 g · m^{-2} (Ajtay et al., 1979). We propose that the sequestering of the organic matter reservoir belowground in grasslands provides these systems with their large ability to adjust and recover from environmental perturbations. At the exteme, the plant community may be degraded or even destroyed, but as long as the soil organic matter pool is not depleted the system will retain its basic fertility and ability to recover.

We hypothesize that semiarid grasslands in the northern Great Plains of North America will be minimally impacted by long-term exposure to *expected* concentrations of SO_2 from energy development because of their structure, dynamics, and adaptations to dry conditions. Additionally, because of their large capacity to recover from perturbations, on those occasions when they are negatively impacted by air pollution, they will recover much more rapidly than other ecosystem types.

References

Ajtay, G. L., P. Ketner, and P. Duvigneaud. 1979. Terrestrial primary production and phytomass. In *The Global Carbon Cycle*, B. Bolin, E. T. Degens, S. Kempe, and P. Ketner, eds. pp. 129–182. New York: Wiley.

Bailey, H. P. 1979. Semiarid climates: Their definition and distribution. In *Agriculture in Semiarid Environments*, A. E. Hall, G. H. Cannell, and H. W. Lawton, eds. pp. 73–97. Ecol. Stud. 34. New York: Springer Verlag.

Durran, D. R., M. J. Meldgin, M. K. Liu, T. Thoem, and D. Henderson. 1979. A study of long range air pollution problems related to coal development in the northern Great Plains. *Atmos. Environ.* 13:104–1037.

FAO. 1979. *1978 FAO Production Yearbook*. Vol. 32. Rome: Food and Agriculture Organization of the United Nations.

Guderian, R., and K. Kueppers. 1980. Response of plant communities to air pollution p. 187–199. In *Effects of Air Pollutants on Mediterranean and Temperate Forest Ecosystems*, P. R. Miller, ed. Berkeley, Calif.: U.S.D.A. Forest Service Gen. Tech. Rep. PSW-43.

Holling, C. S. (ed.). 1978. *Adaptive Environmental Assessment and Management*. International Series on Applied Systems Analysis No. 3. New York: Wiley.

Kozlowski, T. T. 1980. Impacts of air pollution on forest ecosystems. *BioScience* 30:88–93.

Lauenroth, W. K. 1979. Grassland primary production: North American grasslands in perspective. In *Perspectives in Grassland Ecology*, N. R. French, ed. pp. 3–24. Ecological Studies, Vol. 32. New York: Springer-Verlag.

Linzon, S. N. 1978. Effects of airborne sulfur pollutants on plants. In *Sulfur in the Environment: II. Ecological Impacts*, J. O. Nriagu, ed. pp. 109–162. New York: Wiley.

Ludwick, J. D., D. B. Weber, K. B. Olsen, and S. R. Garcia. 1980. Air quality measurements in the coal fired power plant environment of Colstrip, Montant. *Atmos. Environ*. 14:523–532.

Margalef, R. 1975. Diversity, stability and maturity in natural ecosystems. In *Unifying Concepts in Ecology*, W. H. van Dobben and R. H. Lowe-McConnell, eds. pp. 151–160. The Hague: Dr. W. Junk, b.v., Publ.

Miller, P. R., and J. R. McBride. 1975. Effects of air pollutants on forests. In *Responses of Plants to Air Pollution*, J. B. Mudd and T. T. Kozlowski, eds. pp. 196–235. New York: Academic Press.

Odum, E. P., J. T. Finn, and E. H. Franz. 1979. Perturbation theory and the subsidy-stress gradient. *BioScience* 29:349–352.

O'Neill, R. V., and D. E. Reichle. 1980. Dimensions of ecosystem theory. In *Forests: Fresh Perspectives from Ecosystem Analysis*, R. H. Waring, ed. pp. 11–26. Corvallis: Oregon State Univ. Press.

Reinert, R. A., A. S. Heagle, and W. W. Heck. 1975. Plant responses to pollutant combinations. In *Responses of Plants to Air Pollution*, J. B. Mudd and T. T. Kozlowski, eds. pp. 159–177. New York: Academic Press.

Scale, P. R. 1980. Changes in plant communities with distance from an SO_2 source. *Effects of Air Pollutants on Mediterranean and Temperate Forest Ecosystems*, P. R. Miller, ed. pp. 248. Berkeley, Calif: USDA Forest Service Gen. Tech. Rep. PSW-43.

Scheffer, T. C., and G. G. Hedgcock. 1955. *Injury to Northwestern Forest Trees by Sulfur Dioxide from Smelters*. Washington, D.C.: USDA Forest Service Tech. Bull. 1117.

Singh, J. S., W. K. Lauenroth, and D. G. Milchunas. 1983a. Geography of grassland ecosystems. *Progr. Phys. Geogr*. 7:46–80.

Singh, J. S., W. K. Lauenroth, R. K. Heitschmidt, and J. L. Dodd. 1983b. Structural and functional attributes of the vegetation of northern mixed prairie of North America. *Bot. Rev*. 49:117–149.

Smith, W. H. 1981. *Air Pollution and Forests*. New York: Springer-Verlag.

Van Veen, J. A., and E. A. Paul. 1981. Organic carbon dynamics in grassland soils. I. Background information and computer simulation. *Can. J. Soil Sci*. 61:185–201.

Index